WELCOME TO FLYING:
A PRIMER FOR PILOTS

D1403223

WELCOME TO FLYING:
A PRIMER FOR
PILOTS

BY DON DWIGGINS

TAB BOOKS Inc.
BLUE RIDGE SUMMIT, PA. 17214

FIRST EDITION

FIRST PRINTING

Copyright © 1984 by TAB BOOKS Inc.

Printed in the United States of America

Reproduction or publication of the content in any manner, without express permission of the publisher, is prohibited. No liability is assumed with respect to the use of the information herein.

Library of Congress Cataloging in Publication Data

Dwiggins, Don.
    Welcome to flying.

    Includes index.
    1. Airplanes—Piloting. I. Title.
TL710.D85   1984   629.132′52   83-18097
ISBN 0-8306-2362-0 (pbk.)

Cover photograph courtesy of Mooney Aircraft Corporation.

Contents

Introduction

Can I learn to fly an airplane? This question runs through the minds of many people whose only experience with flying is riding as a passenger aboard a jet airliner. To some, the handsome airline captain sitting in the left seat of a crowded cockpit, skillfully monitoring a confusing array of dials and switches, is some kind of Superman.

As a matter of fact, the airline captain *is* a very special kind of flier, but he is only one of hundreds of thousands of Americans who are licensed to fly aircraft. Most are ordinary men and women—doctors, lawyers, businessmen, mechanics, salesmen, housewives, students—who have taken up flying for the sheer pleasure of it.

These folks who share the airman's world have no special qualities that set them apart from other millions of citizens who expertly operate automobiles every day without thinking of themselves as specially gifted. What they do share are such things as good health, average eyesight and hearing, and a normal emotional stability.

No more special skills are required to become a licensed pilot than to become a competent motorist. Just as there are racing drivers, truck drivers, bus drivers, and family car drivers, so there are airmen who learn to operate a vehicle safely in the sky.

Older people, with more mature judgment, sometimes make better pilots than younger men and women because they have learned to exercise caution and thus are ready to respond calmly to an emergency should it arise. For the military pilot, the racing pilot, or the aerobatic pilot, precise timing and muscular coordination are essential, but for general flying, a calm, orderly mind provides the key to good airmanship.

More than three-quarters of a million Americans holding active pilot certificates—student, private, commercial, and airline transport—enjoy flying today. A student pilot is limited in what he can do (he may not carry passengers, for example). Private pilots can carry passengers, though not for hire, in aircraft that they are licensed to fly. Commercial and ATP pilots are professional airmen and so are entitled to charge for their services in aircraft they are rated to fly.

To fly different kinds of aircraft, you must have passed special tests and hold different kinds of ratings—single-engine and multi-engine craft, helicopter, gliders, balloons, seaplanes, etc. As an example, more than 5,000 airmen currently are licensed to fly rotorcraft, and this requires special flight training.

Flying requires no special physical stamina, though in the early days of transport flying women were barred—it was felt that the physical strain of maneuvering a big transport would be beyond their capabilities. One lady pilot, Helen Ritchey, defied the rule to become the first of her sex to get a job flying for Central Airlines, back in 1934. Today you're likely to find a woman in the right or left seat of a transport cockpit doing her thing.

All pilots must, however, be free from any physical defects or medical problems that could incapacitate them suddenly while in flight or could prevent them from flying an aircraft safely under all expected conditions. To keep a close surveillance over the physical fitness of airmen, the Federal Aviation Administration (FAA) has designated thousands of physicians as Aviation Medical Examiners, who conduct periodic physical examinations (at least every two years for private pilots).

Such examinations are thorough and are good life insurance, as they may detect some minor health problem that could become serious if neglected. Commercial pilots who fly for hire must be examined annually, and ATP pilots every six months.

Medical conditions that would bar a citizen from becoming a pilot are about the same as those which would disqualify him from driving a car on a highway—certain types of heart disease, chronic alcoholism, drug addiction, or serious mental disorders. Waivers and exemptions may be obtained in some cases.

Again, just as a motorist faces suspension or revocation of a driving license for serious infractions of vehicular codes, so an airman who violates the Federal Aviation Regulations (FAR) may have his ticket suspended or revoked.

Assuming, however, that you are a normal citizen, mentally and physically fit to drive a car, to become a student pilot you have only to visit a designated FAA medical examiner, pass a routine physical examination for a student pilot license (a Third Class Medical), be 16 years old (14 for glider pilots), and be able to read, write, and understand the English language.

Once you have shown that you are familiar with certain air traffic rules by passing a simple written examination, you will be on your way to that first solo flight—that great moment when your Certificated Flight Instructor (CFI) feels you're ready and endorses your Student Pilot Certificate to that effect. It all sounds simple, and in fact it is.

Personal flying has grown tremendously since World War II, and though it suffered a setback along with other segments of travel activity during the recent recession, things are returning to normal today. Much of this current enthusiasm for flying stems from introduction of relatively inexpensive aircraft called *ultralights*. You can buy or build one for only a few thousand dollars, and as of now no pilot's license is required. Once you've learned to handle the little craft, you may want to step up and fly conventional airplanes (which, of course, *do* require an FAA airman's certificate to operate).

This book is not meant to be a technical flight instruction manual or even a detailed examination of the complex fields of meteorology, navigation, communications, or flight dynamics. You'll cover all these in the classroom when you elect to start flight training. Rather, it is meant to be an orientation for the non-pilot and the beginning pilot alike, as well as a refresher course for the licensed airman. This book is meant to help you get the maximum enjoyment from owning and/or operating an airplane in a safe manner.

I hope to pass along to you some basic tips on what to do and what not to do, based on more than forty years of flying and instructing experience. I hope that this will help you become a safe and cognizant pilot through a fuller awareness of what *airmanship* really means.

Chapter 1
The Open Skyways

We are accustomed to living in a two-dimensional world, one with breadth and depth—"room to move around in," we say—but relatively little height. True, we live in comfort from sea level to altitudes of more than a mile high, in cities like New York, Denver, or Los Angeles, but in our everyday lives we're used to moving around on one plane—if you exclude elevators!.

Above us, however, lies a vast ocean of air called the *atmosphere*, extending from the surface upward hundreds of miles. The air at the bottom of this ocean is denser, as it supports all the air above it, and is relatively warm, averaging 59 degrees Fahrenheit the year around in the United States.

As you go higher, the temperature becomes lower at what is called a *normal lapse rate* of about 3½°F for each 1000 feet, until finally at an altitude of seven miles it stabilizes at near 67°F below zero. This upper region, the *stratosphere*, was so named because its temperature changes very little.

We are constantly learning more about this high layer of sky, where jet stream activity, cosmic radiation, and solar winds produce an immense effect on world weather patterns. But our concern as private pilots is not the stratosphere but the lower layer called the *troposphere*, so named because it is ever-changing—a deep sea of swirling currents, tides, and whirlpools of hot and cold winds and vapors that constitute our "weather."

It may be difficult to realize that there is a tremendous weight involved—at sea level, some 20 tons of pressure is exerted on the body of the average person. This pressure naturally diminishes as we fly higher, and at 18,000 feet MSL (Mean Sea Level), it's half that at sea level. This means, of course, that for every breath of air we inhale at 18,000 feet, our lungs get only 50% as much oxygen as at sea level.

The troposphere and stratosphere in which most aircraft fly are divided into layered corridors of navigable or useful airspace with a ceiling considered to be at 75,000 feet, or Flight Level 75 (each 1000 feet constitutes a separate Flight Level). Navigable airspace reaches to the ground at airports, and generally follows the Earth's contours clear of mountains, tall towers, or other obstructions to flight.

Highways in the Sky

This open sky is alive with ever-present signals from radio navigation facilities. These signals define the roughly 250,000 miles that make up our Federal Airways System.

This remarkable sky network is in many ways similar to our common surface freeways and highways. There are no charges currently levied to fly them in the United States, though an "airway tax" has been proposed from time to time to maintain the system with user fees. As of now they are free and available for all to use—skyways that measure eight miles wide! Across the border in Mexico, a private organization called RAMSA operates their air navigation facilities, and pilots flying in that country must pay a user fee. In the good old U.S.A., the only such fees you may have to pay are the landing fees charged at some of the busier terminals, and frequently, overnight parking charges.

These freeways of the sky are complete with aerial versions of signs, access roads, directional guides, and even "parking" areas—holding patterns where aircraft wait their turns to land at busy terminals. Airplanes assigned to holding patterns fly "race-track" ovals at assigned altitudes while waiting to descend and land.

You can't *see* these aerial freeways; they are invisible except on charts, where they appear in relation to the terrain beneath them. Rivers, lakes, highways, large and small cities, towns, and villages, mountains, bluffs, cliffs, ridges, swamps, and marshes are all plainly marked.

Aeronautical charts also indicate minimum safe altitudes to fly, as well as navigation aids (navaids), electronic devices that provide guidance and location data to airmen. These range from small location marker beacons to vast, long-range surveillance radar systems linked by microwave to traffic control facilities hundreds of miles distant.

The basic navaid, the backbone of the electronic airways system today, is called the *VOR*. It is a very high frequency navigational facility providing 360 separate courses to or from the station. It is identified in the pilot's headset by a three-letter Morse code, or by a recorded voice. There are roughly one thousand such stations to assist you in flying across the United States; you simply tune them in one at a time with your receiver. Some VOR frequencies offer weather and other important information while you are flying toward or away from a station.

While there are many off-airways areas of airspace where you may fly today without radio contact, all air traffic is in fact similar to automobile traffic in one respect: It cannot operate helter-skelter, but must conform to established regulations, either VFR (Visual Flight Rules) or IFR (Instrument Flight Rules).

Generally speaking, a pilot flying by IFR navigates primarily by instruments, relying on them and on radio instructions from air traffic controllers in order to stay safely separated from other aircraft.

A pilot flying by VFR, however, relies on his own eyes to guide his airplane in relation to terrain landmarks and to stay clear of other traffic by a system called "see and be seen." (If you're flying an ultralight aircraft, however, with or without a pilot's license, you are not permitted to fly the federal airways or operate from a controlled airport, except under a special waiver.)

IFR operations are employed if you are flying in weather below VFR minimum ceiling (1000 feet) and visibility (one mile) parameters outside of controlled airspace. To qualify for IFR flying, a civilian pilot must first pass an instrument flight test and receive a special instrument rating from the FAA. He must also operate under strict IFR regulations when flying in "positive control" airspace, generally above 24,000 feet.

All air traffic in the United States operates under a "common system," with both civilian and military aircraft controlled by identical procedures from the same facilities. Traffic within a 3—30 mile radius of an airport is controlled by either a military control tower or one of the more than 350 FAA airport control towers.

En route traffic on federal airways is controlled by a score of Air Route Traffic Control Centers, located strategically throughout the nation. Each center and tower handles traffic within its own

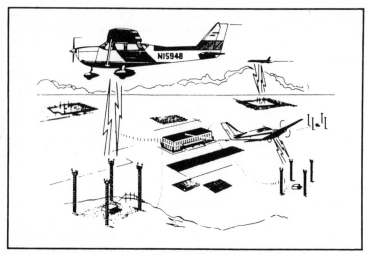

Fig. 1-1. Remote communication sites transfer your air-to-ground radio call to Air Route Traffic Control Center. (courtesy FAA)

area, using radar and communications equipment to keep traffic flowing smoothly and safely.

As you fly from one VOR to another on a cross-country run, control of your flight is transferred from center to center, and finally from center to tower as you approach your destination to land (Fig. 1-1).

A third air traffic facility, the Flight Service Station (FSS), devotes most of its attention to the hundreds of thousands of non-airline civilian pilots who make up the bulk of the general aviation airman population, plus a growing number of military pilots. You'll find more than 300 FSS and combined FSS/Towers scattered across the nation, each serving an area of roughly 400 square miles. In 1983, the FAA's National Airspace System Plan launched a program to consolidate its network of FSS units into an integrated system of 61 automated facilities, providing FSS specialists with a faster means of computerized information retrieval, better to brief pilots on weather and other flight conditions.

Today's FSS specialists are knowledgeable about their areas' local terrain problems, and offer expert advice during preflight or inflight briefings on weather, the best route to fly, best altitudes and winds, and other data that may make your cross-country flight safer and more enjoyable.

At many FSS stations, emergency very high and ultra high frequency (VHF/UHF) direction-finding equipment helps lost pilots by giving orientation instructions called Direction-Finding (DF) Steers. By simply calling the FSS for assistance, a lost pilot is given a compass heading to fly to reach his destination most advantageously.

All these remarkable electronic aids in airport towers, en route traffic centers, and Flight Service Stations are linked by more than 500,000 miles of land lines to maintain an orderly, safe traffic flow across the sky. The FAA publishes special manuals with complete information on these facilities and how to use them. The best of these is the *Airman's Information Manual,* published periodically with Basic Flight Information and ATC Procedures.

If all this sounds impressive, it is. The Federal Airways System is something to be proud of, although it faces a constant struggle to meet growing traffic demands. By 1992, the FAA estimates, general aviation aircraft will be flying more than 64,000,000 hours per year, with some 315,000 planes and 1,099,300 licensed pilots involved. The FSS workload by 1992, FAA believes, will total some 10,000,000 contacts between general aviation aircraft and Flight Service Stations and combined Station/Towers.

In earlier days of flying, communications and radio navigation were not as important as they are today. There were far fewer aircraft, and navaids were not so sophisticated. But modern airmen quickly learn to use their radios to fullest advantage, as communications is now an integral part of the airman's world.

Beginning pilots may at first seem frightened by their radios, yet as one FAA examiner put it, "these same people feel perfectly at ease talking on the telephone." There's really not much difference, once you learn to use the system to advantage.

To make the vast ATC system work smoothly, the FAA operates one of the world's largest networks of teletypewriter, telephone, and radio communications. Its operational weather service alone consists of some 100,000 channel miles of teletypewriter circuits. Another 360,000 miles of telephone circuits link FAA facilities, airline flight dispatch offices, military flight service centers, air defense units, and search and rescue organizations.

In the latter category, a novel new international cooperative effort called SARSAT (Search and Rescue Satellites) was inaugurated in 1982 to speed help to aircraft in distress. Should an aircraft crash in a remote area, an on-board Emergency Locator Transmitter (ELT) is automatically activated, sending out a line-of-sight signal. Both U.S. and Russian satellites are now equipped to pick up such distress signals and pinpoint the source with amazing precision. Search and rescue planes are then dispatched to locate any possible survivors. Other satellites like the military's NAVSTAR further implement normal navigation procedures as we enter the "Space Age" of personal flying.

So there you have it—the open sky, thousands of feet of layered traffic zones through which aircraft may fly in orderly fashion, often under radar guidance, to reach their destinations safely. Despite the enormity of the growing airline traffic load, the biggest user of these invisible skyways remains general aviation. Simply by filing a flight plan—either VFR or IFR—a pilot knows his flight is being closely watched by trained experts whose lives are dedicated to safe flight.

Obviously, there are vast reaches of uncontrolled airspace—below 1200 feet, off airways, and in assigned areas—where the student pilot can try out his new-found wings and learn the fundamentals of flight in comfort and safety before setting off on long-distance trips along the airways. Your local airport operator can show you where these local practice areas are located. It is there, at first with your flight instructor along, that you will win your wings by proving to a designated FAA Flight Examiner that you are indeed a qualified pilot.

Room for Everyone

There's really room for virtually everybody in the sky because of this "open-sky" philosophy designed to permit intermingling of many types of aircraft operating at different airspeeds, through careful air traffic control.

The question now arises—what kind of aircraft do you want to fly? On the used-plane market you can find a wide range of somewhat inexpensive "time-builders" available in which to practice your maneuvers. There also are new two- and four-place trainers and utility category aircraft you can buy and fly for business and/or pleasure, to suit your budget. But whether you decide to start off flying unlicensed ultralights (that weight under 254 pounds empty, carry no more than five gallons of fuel, and can't fly faster than 55 knots), or conventional aircraft, you will want to master a few simple but hard and fast rules to become a good pilot.

A Look at the Future

Good as it is, the National Airspace system currently is being upgraded in facilities and equipment to meet projected needs and demands of aviation safety by the year 2000. A National Airspace

6

System Plan adopted early in 1982 has a number of goals to meet, requiring expensive changes in electronics and other systems, paid for largely by aviation fuel taxes of five cents a gallon.

- ☐ Ground-based air traffic control services will continue to be the fundamental means for assuring aircraft separation in flight.
- ☐ The FAA communications system will be improved and better integrated.
- ☐ The current Instrument Landing System (ILS) will eventually be replaced by the newer Microwave Landing System (MLS).
- ☐ Work will continue on airport improvement to meet demands of heavier traffic.
- ☐ Automated and semi-automated weather observation systems will speed pilot briefings at both manned and unmanned facilitites.
- ☐ Direct pilot access to real-time weather reports will be provided through improvements to the FAA's Automated Flight Service Station System.
- ☐ Existing en route and terminal air traffic control computers will be replaced with a modern, more reliable system capable of supporting higher levels of automation.
- ☐ FAA will proceed with upgrading terminal automation systems, ARTS II and III, while displays and software are being developed for terminal operations.
- ☐ Quality and efficiency of providing weather and flight plan service to all users will be improved, using the Automated Flight Service Station program.
- ☐ Modernization of air traffic control management will be provided between national en route and terminal flow traffic handling.
- ☐ An airborne collision avoidance capability independent of ground control will be implemented, called the Traffic Alert and Collision Avoidance System (TCAS). It will incorporate a new altitude-reporting Mode S transponder; replacing the current Air Traffic Control Radar Beacon System.
- ☐ En route primary radar will be phased out in the year 2000, though terminal primary radar with a weather channel will be continued.
- ☐ Improved weather services will be provided for the air traffic control system by a new weather radar.

With this growing complex of air traffic control systems, flying will become increasingly safe as the volume of air traffic grows over the next two decades.

Chapter 2

Your First Airplane

Buying an airplane is a little more like buying a boat than a car. Most automobile shoppers look primarily for cheap, fast, and comfortable transportation. Boat and airplane owners add an extra requirement—*fun*.

The FAA made an in-depth study of who flies what, and noted that "in the overall economy, major changes are marking trends in general aviation. In recent years, general aviation has become increasingly important as a means of transportation for business use . . . Orders for turboprop and jet aircraft are approaching backlogs of three years, while in contrast, some of the smaller aircraft production was suspended for varying periods early in the 1980s."

General aviation, says FAA, is severely affected by changes in lifestyle: "New aircraft are required to be more fuel-efficient and all aircraft must be equipped with elaborate new safety equipment."

Though business flying today accounts for the most time spent in the air, personal flying now runs a close second.

When you begin flying as a student pilot, your logged time, of course, is purely instructional (dual) time, so whatever your goal as a pilot, let's now consider what kind of airplane you'd want to buy to learn to fly in as an alternative to flying in a school-owned aircraft.

There's an old story of the would-be boat owner who visited an exclusive yacht showroom and asked the salesman, "Hey, how much is that yacht over there?"

"If you have to ask," the salesman replied haughtily, "you can't afford it!"

We've all watched with admiration (and some envy) the pilot of a $200,000 executive twin-engine aircraft climb into his 275-mph job and zoom off for Acapulco or London or Hong Kong. But let's buy that big one tomorrow—today we're looking for something more in the "rowboat" category before graduating to those sleek flying yachts.

Time-Builders

It makes little difference when you start out how big or fast your airplane is, for you can get just as much enjoyment and actually learn more fundamentals of flying in smaller, less-complicated aircraft.

In fact, it may be well to start off with a second-hand trainer costing under $5000. Such ships are called "time builders" because they serve just that purpose—to help you gain flying experience and log the time necessary to qualify for your pilot rating.

Your first Airman's Certificate—your Student Pilot Rating—is issued by an FAA Medical Examiner, and no prior logged flight time is required. This certificate is endorsed by your instructor when he deems you ready to fly solo. From that point, you are eligible to fly alone, practicing assigned maneuvers, until you've met the requirements that qualify you to take your FAA flight test for your Private Pilot Certificate.

An applicant for a Private ticket must have logged at least 40 hours, including 20 hours of dual instruction time from a CFI and 20 hours of solo flying. Of the latter, at least 10 must be in an airplane (as opposed to a simulator) and another 10 hours must be flying cross-country. You must have landed at an airport more than 50 miles distant from your home base on each cross-country flight, and made at least one 300-mile cross-country flight, landing at three airports each more than 100 miles distant from the other. In addition, you must have made three solo takeoffs and full-stop landings at an airport with an operating control tower.

After completing your cross-country flying, you must take at least another three hours of dual from a CFI, including a review of procedures and maneuvers previously learned plus additional instruction to prepare you for your flight test. The FAA also requires you to have logged instruction in primary maneuvers flown solely by reference to instruments, including climbs and descents using radio aids or radar directives.

This extra emphasis on IFR flying is a relatively new idea, and a good one. Inevitably you will run into unexpected bad weather, and a basic understanding of IFR techniques, in controlling the aircraft solely by reference to instruments until the emergency passes, could well save your life.

A knowledge of flying by reference to instruments rather than the natural horizon also helps you to understand many basic aerodynamic principles. The gauges are your extra "eyes," helping you sense what your aircraft is doing. There was a time when CFIs intentionally covered the airspeed indicator and taught their students to fly "by the seat of their pants." There was a valid reason for this—the kinesthetic muscular sense really does give you information, but, sadly, it can be wrong!

True enough, an awareness of the sound of the wind rushing past windows and around struts and wires, and the feel of the controls (hard in fast flight, mushy in slow flight) tell a lot about what your plane is doing. Today, however, airplanes fly much faster, carry engines of more power, and have wing loadings and stall speeds more than double those of early civilian trainers. Hence, "seat of the pants" flying is not enough. To be a good pilot, a basic knowledge of IFR techniques is indispensible. Current FAA flight examinations, in fact, have been changed accordingly, to

Fig. 2-1. Thousands of pilots learned to fly in the graceful old Piper J-3 Cub.

Fig. 2-2. The author spent many happy hours flying this Luscombe 8E Silvaire to distant desert airfields.

qualify you to handle the hotter ships you'll probably by flying in the future.

Regardless of the high-performance capabilities of modern aircraft, a 40-year-old Piper J-3 Cub (Fig. 2-1) can't be beat for getting you started properly on flight fundamentals. The Luscombe Silvaire (Fig. 2-2) and the Cessna 120/140 (Fig. 2-3), both of which appeared on the market just after World War II, make equally good trainers. The Cub is a two-place tandem-seating craft, while the Luscombe and Cessna 120/140 are two-place side-by-side seaters.

To shop around for a used trainer, visit your local airport, study the newspaper classified ads, or pick up a copy of the *Aircraft Bluebook* (from the Aircraft Dealers Service Association, P.O. Box 621, Aurora, CO, 80010) which lists current wholesale and retail values and factory list prices on all models. Also check out several aircraft shopping guides such as *Trade-A-Plane*.

A newer, popular four-place aircraft widely used in pilot training is the Piper Cherokee PA-28 150, powered with a 150-hp Lycoming engine. It first appeared in 1961 for just under $10,000. Today they'll run anywhere from $6250 to more than $20,000, depending on condition and equipment installed.

Buying your first airplane is an event not unlike going into marriage, it's been said: It should be done as if your whole happiness and very life depend on it. Whatever price you decide on, your decision on make and model is only the start of a selection process. This process should not be sacrificed to compulsive buying.

Just as in purchasing a home, a stereo set, or a fine, expensive camera, there are technical details to investigate in order to judge the quality, performance, and enjoyability you'll get for your money. It's a good idea to talk with an expert first; you cannot be expected to base a wise decision on your own background experience unless you've been around airplanes for a while.

Once you have a rough idea of how much you want to spend, or can afford, your next step is to make up a list of things you expect to get in that price range. You'll have a choice between two types of landing gear—the conventional taildragger or the tricycle gear (Fig. 2-4). There are high-wing trainers (Pipers and Luscombes) and low-wingers like the Piper Cherokees, etc. Some come with tandem seating; others have side-by-side seating—more chummy, but presenting a different perspective in turns with the horizon higher one way and lower the other.

When buying a new plane, after comparing costs, carrying capacity, performance, and other criteria, you'd be wise to go see a reputable dealer who can't afford to sell you a dog. If you buy from a

Fig. 2-3. The Cessna 140 side-by-side two-seater is an excellent trainer.

Fig. 2-4. Cessna 150s are popular beginning trainers today.

local dealer, you get a factory warranty that guarantees replacement parts and service when and if required.

An equally important factor today is the kind of radio equipment installed—which could cost even more than the basic airplane itself!

Kicking the Tires

Other things to explore in buying a used plane are the number of hours the engine has been run since new or since its last major overhaul (SMOH), quality of the paint job, condition of fabric covering or metal skin, and condition of interior upholstery—all expensive items.

A second-hand airplane may appear like new, but may also have hidden mechanical problems waiting to give you a real headache later on. To avoid this, ask the salesman to have an "evaluation check" run by a neutral aviation service station. Have a mechanic pull the oil screen and check for metal particles that may indicate undue engine wear. Better still, have him run a *comparator check*, testing the compression of each engine cylinder. If there's a hidden pressure bleedoff, find out where the air is going.

For example, a hissing in the carburetor may indicate a burned intake valve, while a hissing noise in the exhaust, when you pull the propeller through, can mean a worn exhaust valve. Air escaping through the crankcase, audible through the oil filler spout, can mean worn piston rings.

Kicking the tires may not tell you much, except to draw attention to the landing gear. If the plane you're considering has retractable gear, have the salesman demonstrate a retraction check, either in flight or on a test stand. Retractable gear, along with engine and radio equipment, is an expensive item.

While looking over the tires, check them for any abnormal wear that could mean misalignment of the wheels, scissors assembly, or axles. Unbalanced wheels will show uneven wear that can induce dangerous vibrations, which in turn can set up vibrations damaging to other parts of the aircraft. Parking in puddles of gasoline or oil can cause rapid deterioration of tire rubber. Look also for cuts, bruises, and foreign objects embedded in the tires.

If the plane you want to buy has oleo-type shock absorbers, check them for cleanliness, leaks, cracks, and possible bottoming of the pistons (Fig. 2-5). Shock cords, when used, should be free from wear, stretching, or deterioration. Generally, spring-steel gear require little if any maintenance, but they should be checked for cracks in the area where attachment brackets are riveted to the fuselage. Wrinkles in the fuselage skin are sure danger signs, indicating past hard landings.

In shopping for a ragwing (fabric-covered) aircraft, be sure you

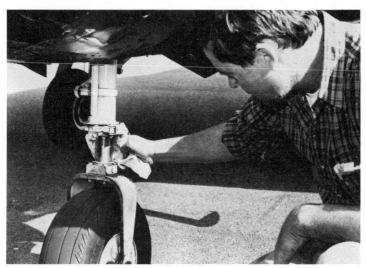

Fig. 2-5. Preventive maintenance is a pilot's responsibility. Here he checks nosewheel oleo strut for cleanliness.

Fig. 2-6. Exposed pitot tube should be checked for any bugs.

have a punch test run to determine the fabric's condition. Aircraft parked outside, exposed to sun, rain, smog, and salt air are more susceptible to deterioration than aircraft that have been regularly hangared. Even metal aircraft suffer when parked outside and exposed to acid rain conditions, incidentally; look for telltale corrosion streaks under the wings, where the rain runoff dribbles down (Figs. 2-6, 2-7).

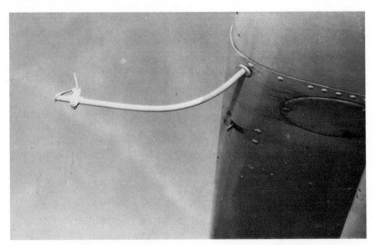

Fig. 2-7. Some pitot tubes, like this one on a Cessna 140, have a special flap that covers opening when not in flight.

Wooden propellers should be inspected for lamination separation and uneven balance, which may result from leaving the plane parked with the prop vertical, so that moisture may seep to one end. Metal props sometimes are deliberately left vertical to prevent birds from roosting on them. They should be checked for leading edge nicks caused by foreign objects sucked up into the spinning blades during stationary runups.

Under the hood (in addition to the standard evaluation check), look for any vibration damage to exhaust and heater manifolds that could admit poisonous carbon monoxide fumes to the passenger cabin. Electrical wiring should be closely examined for deterioration by wear or friction fuel strainer screens checked for water or dirt contamination, and oil lines inspected for leaks or decomposition (Figs. 2-8 through 2-11).

A thorough pre-purchase aircraft inspection may include items normally attended to only during routine daily preflight checks. By so doing you can learn a lot about the maintenance habits of the former owner. Corroded battery terminals, oil accumulations beneath the engine cowling or on the undersurface of the fuselage, cracked or crazed plexiglass windshields, and the general condition of the cabin interior are other revealing tips on whether the former pilot/owner attended to preventive maintenance properly or let things go until time to sell. Don't let a fancy paint job fool you.

Next, look back inside the fuselage tailcone of a metal aircraft

Fig. 2-8. A careful pilot makes sure all hoses are tightened.

Fig. 2-9. Routine inspection of spark plugs shows engine condition.

for any sign of corrosion, and check to make sure all electrical wiring and control cables run free and clear of each other. Check the seat belt and shoulder harness fastenings, the play in the control column bearing and rudder pedal bushings, and the condition and tension of all control cables, push-pull rods, and pulleys (Figs. 2-12, 2-13).

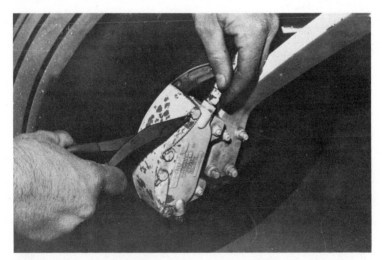

Fig. 2-10. Safety wires are routinely checked. Make sure they're installed so tension is against untightening.

Fig. 2-11. Make sure spark plug lead wire attachment nuts are tight.

See that all flight and engine instruments are reading properly, or you may have them in the shop for expensive overhaul after your first flight. Make sure that vacuum-driven flight instruments are not noisy, a sign of worn gyro bearings. A glance underneath the

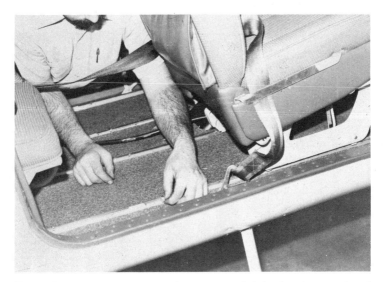

Fig. 2-12. Inspect seat tracks to make sure stop pin is in place to prevent seat from coming out.

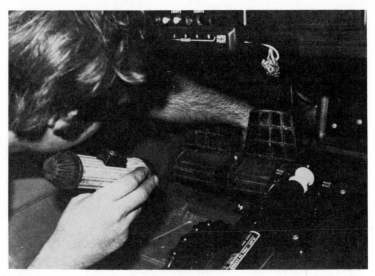
Fig. 2-13. Use flashlight to check hard-to-get-at places like master brake cylinder.

instrument panel can tell you much about how well this important nerve center of the craft has been maintained.

In checking radio equipment, remember that there are two basic reasons for its being there—communications and navigation. If you're buying a used time-builder, you may not need expensive nav equipment, though NAV/COMM sets are in common use.

Normal airport flying requires only a two-way VHF transceiver and a simple VOR radio to work omnirange stations on cross-country flights. However, if you intend to be flying IFR, you'll need a wider selection of radio equipment. Make sure there's enough room on the instrument panel (or elsewhere in the cockpit area) to mount extra electronic gear.

Aircraft Paperwork

One of the most important items to check when buying a used plane are the aircraft and engine logs. These legally must be kept up-to-date by the aircraft owner, who writes down a report on any major repair or alteration as well as preventive maintenance. Periodic inspections are listed there, signed off by licensed Aircraft & Powerplant mechanics (A&Ps). If all repairs, installations, and changes of parts are properly noted and signed off, chances are you're buying an aircraft that has been carefully maintained. It should give you expected good service for its age and condition.

When you decide to buy, remember there are some other details to consider to complete the transaction in a businesslike manner. First, make sure you're buying clear title to the aircraft, free of such encumberances as leins, chattel mortgages, or other unsatisfied claims. To accomplish this, either search the aircraft records yourself or let an attorney or an aircraft title-search firm handle it. All aircraft public records maintained by the FAA are on file at the Aircraft Registration Branch, FS-965, FAA Building, 5300 South Portland Avenue, Oklahoma City, OK 73119.

Whenever buying either a used or a new aircraft, you should receive the following documents:

☐ A bill of sale.
☐ An Airworthiness Certificate (FAA Form 1362B).
☐ All logbooks, aircraft, and engine records.
☐ An equipment list.
☐ A weight and balance data sheet.
☐ A maintenance manual, service letters, bulletins, etc.
☐ An airplane flight manual or operating limitation list.

A current 100-hour or periodic inspection certification doesn't mean that the aircraft is in top shape—it only signifies that the A&P who signed it off found it was airworthy *at that time, in his opinion.*

There are certain responsibilities attendant to owning a plane, just as there are in owning a car. These include:

☐ Displaying Registration and Airworthiness Certificates in your aircraft.
☐ Maintaining your ship in an airworthy condition.
☐ Assuring that all maintenance is properly recorded.
☐ Keeping abreast of current operation and maintenance regulations.
☐ Notifying the FAA's Aircraft Registration Branch immediately of any permanent change of address, or sale or export of your aircraft.

To keep your Airworthiness Certificate current, you must have the aircraft inspected by a qualified mechanic at least once every twelve months.

Before you fly an aircraft you have just purchased, you must first apply for a Certificate of Registration. An aircraft is eligible for U.S. registration only if it is owned by a United States citizen and is not registered under the laws of any foreign country.

The FAA Registration Form 8050-1 contains three parts in

duplicate: Part A is the Certificate of Registration; Part B is the Application for Registration; Part C is the Bill of Sale. After filling out the paperwork, send them off to the FAA Aircraft Registration Branch in Oklahoma City, retaining the yellow carbon copy of Part B (the Application for Registration) to display in the aircraft until the FAA returns the Certificate of Registration (Part A). This way, you maintain an unbroken chain of title for your own protection.

The Certificate of Airworthiness (FAA Form 1362B) is issued by an FAA representative after a thorough inspection that establishes the aircraft as safe to fly. This certificate remains effective as long as the ship is properly maintained and operated within its specified limitations. It's your best insurance against accidents due to mechanical problems.

While the FAA requires a periodic inspection at least once a year by an authorized A&P mechanic, experience shows that preventive maintenance inspections should be done every 25 hours of flight time or less, and minor maintenance every 100 hours. New regulations require that all preventive maintenance inspections and repairs must be entered in the aircraft and engine logs and signed off by the plane owner.

An aircraft used to carry passengers for hire, or for flight instruction for hire, must be inspected every 100 hours by a certified A&P mechanic or other authorized expert. You can run your own daily and preflight inspections, and it remains the pilot's responsibility to perform a thorough inspection prior to each flight (Fig. 2-14).

Certificated pilots are permitted to perform line maintenance, preservation, and upkeep (preventive maintenance) on aircraft owned or operated by them. This may mean changing a worn tire, painting underneath the engine cowling with corrosion-checking zinc chromate, changing the crankcase oil or spark plugs, and other routine chores such as you might attend to on your car. Major or minor repairs (such as brazing a cracked nut, recovering a fabric wing, or reboring engine cylinders) must be done or approved by a licensed mechanic.

Many lightplane owners prefer to do their own minor repair work, under the supervision and approval of a licensed mechanic, and on weekends around any small airport you'll find careful owner/pilots enjoying themselves at such tasks. To assist you in this regard, TAB Book#2305 *Inspecting Your Own Plane* is recommended.

Once you've picked out the plane you want to buy, looked it

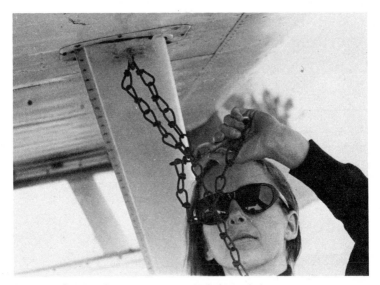
Fig. 2-14. Quick and easy way to secure tiedown chain.

over, completed the transaction, and done all your paperwork, you're now ready to blast off into the wild blue yonder and get acquainted with your new ship in its proper element! You can make the first flight yourself if you're a licensed airman. Better still, if it can be arranged, is to have the prior owner or a CFI take you up to help familiarize you with its controls and handling.

Most airplane owners, in fact, learn to fly before buying their first aircraft, either from an individual Certified Flight Instructor (CFI) or from an approved flying school. You can also take legal instruction from a licensed private pilot, although such time may not be logged toward the number of hours required to obtain your pilot's license.

In the next chapter, we'll discuss basic flight instruction, to help you get started properly. Here's where the fun begins! It may seem hard work at first, but quickly will come a new sense of freedom, enjoyment, and responsibility as a student pilot.

Chapter 3
Where Do I Go To Learn To Fly?

If you want to learn to drive a car, you have several options. You can take instruction from a private individual holding a driver's license, or from a public school offering driver instruction, or from any of numerous commercial schools listed in the yellow pages of your telephone book under Driving Instruction. Once you are judged ready for your road test and have successfully passed a written examination on traffic rules, you demonstrate your skill and judgment on the road. If you pass the test, you are ready to join the traffic stream on our crowded highways as a safe driver.

In the same way, there are individual Certified Flight Instructors (CFIs) and schools licensed by the federal government to give flight instruction for hire and sign your student pilot's license signifying that (in their judgment) you are ready to venture away from the airport alone.

Flight Training Quality

Today there is an organization called the National Association of Flight Instructors (NAFI) which operates under a Code of Ethics specifying that they will provide the best possible instruction according to accepted techniques and standards. A NAFI member performs his professional services always for the good of the student; he keeps confidential the student-instructor relationship, releasing information concerning the student only as it benefits the student, serves professional purposes, or is required by law. Because the Federal Aviation Regulations are constantly being revised and upgraded to conform to changing times, you would do well to know if your instructor is a NAFI member.

In addition to the independent CFIs who give instruction on a part-time basis (usually on weekends), there are many fine aviation courses in state high schools and universities that offer flight instruction as an extension of their curricula. An outstanding school of this type is conducted by the Ohio State University's Department of Aviation at Columbus, Ohio.

By far the largest number of flight and ground training schools are privately operated. The FAA publishes a List of Certificated Pilot Flight and Ground Schools, listing more than 1,000 such training bases throughout the 50 states and abroad (Advisory Circular No. 140-2D). In this publication (available from the Department of Transportation, Federal Aviation Administration, Distribution Unit, TAD-484.3, Washington, D.C., 20590), the certificated schools are code-listed as to their services—ground or flight training, in airplanes, gliders or helicopters, and whether their courses are primary, basic, or advanced.

Many of these schools maintain their own Examining Authority—a designated representative of the FAA licensed to present graduates for certification without further test by the FAA.

Because of this emphasis on the quality of flight training, maintained by frequent supervisory flight checks of CFIs by FAA inspectors, the chances are slim that you will encounter a mentor who is only interested in training you to pass a flight check. Dedicated flight instructors know that your future safety and enjoyment of flying depend on you learning a thing called *airmanship,* which is the sum total of your ability to feel at home in your new environment, the sky, and prepare yourself to meet emergencies intelligently and quickly.

Pre-Solo Instruction

There is no minimum number of hours of dual instruction required before your instructor may deem you ready to solo. I remember well one man of 45 who had worked on airplanes professionally as an engineer for many years; he was ready to solo at the end of the second hour. However, skillful as he was, there was much to teach him about how to stay out of trouble and what to do if trouble should occur. (He soloed at the end of the sixth hour.)

Other students, slower to learn or unable to keep flying on a firm schedule, require more than the "average" eight hours of instruction before solo. While there may be a psychological hand-icap to logging more than the "average" dual time, it is unwise to attempt to fly solo before you have mastered the basics of takeoff, climb, cruise in level flight, turns, approach, and landings. Of course you must also know how to taxi on the ground, in no wind and in stiff breezes, both "down the runway" and in crosswinds.

You will want to practice landings in the sky before trying them on the ground, simulating them by performing different kinds of stalls. These may be power-off full-stall maneuvers that simulate actual normal landings of taildragger and tricycle gear aircraft alike. You will also learn to accomplish takeoff and departure stalls and approach and landing stalls, designed to familiarize you with the operating limits of your airplane during these critical maneuvers.

Pre-solo instruction will include much slow flying, maintaining altitude and compass heading control at reduced engine power. This is a condition that you will frequently encounter in normal airport pattern flying, pacing your speed to that of other traffic. Slow flight also teaches you some important basics of how an airplane flies and what the throttle and elevator controls are for.

A Common Confusion

It is surprising that many pilots confuse the function of the throttle and elevator controls in their minds—and sometimes dangerously, in actual practice. This occurs because it seems per-fectly logical that the airplane's speed control should be the engine throttle, just as it is in a car.

Applying power will eventually produce higher speeds, but in the precise art of flying an airplane—particularly in the lower speed regimes near a stall—it is the *elevator* that controls speed. The throttle is used for *altitude* control.

Think this over a moment: You are slow-flying through an airport traffic pattern, heading downwind behind another ship. Sud-denly you feel a warning shudder brought on by an approaching wing stall as the smooth flow of air over the top of the wing breaks away and sets up turbulent eddys that destroy lift.

When this happens, it simply means that the air flowing over the wing is coming at it from below and at too large an angle to flow smoothly over it, causing the stall. This angle of the *relative wind* to the wing's *chord line* (a line from leading edge to trailing edge) is called the *angle of attack*. A full understanding of it is imperative to good airmanship.

When such a situation occurs—when you sense a stall

beginning—the correction is obvious. You must quickly reduce the angle of attack so that the air will again flow smoothly over the top of the wing and produce lift once more.

To do this, the control stick (or wheel) is moved forward sharply, depressing the elevator surfaces into the slipstream. This brings the tail up, depressing the nose and aligning the wing into the relative wind (or direction of flight, whichever you prefer). Doing this reduces the turbulence behind the wing, which we call *induced drag*. Reducing drag lets the airplane move faster through the air. It is obvious that the higher the angle of attack, the more drag on the wing, and the less speed. So to go faster, reduce the angle of attack with the elevator.

You can do this, of course, with the engine idling, but if you shove the stick forward to pick up speed without touching the throttle, you'll simply be going faster, downhill in a shallow dive. In order to maintain altitude at any specific airspeed, you therefore apply more power.

In reality, elevator and throttle work together in maintaining altitude and airspeed within the desired parameters. A good pilot knows how to use these controls simultaneously, as in leveling out from a climb. By moving the stick forward, he begins to increase airspeed from climbing to cruising speed; a careful adjustment of the throttle holds his ship at the preselected altitude at a given airspeed. This is called *stick-and-throttle coordination.*

Let's look at another situation. You are on a final glide approach to land when perhaps 100 feet above the ground you feel your wing shudder. A stall is coming on! You must do something fast. If you let the stall progress, the wing will lose lift to the point where the nose of the airplane will drop down into a dive.

Here, the untrained pilot may well panic and do exactly the wrong thing—pull back on the stick to bring the nose up with the elevators. What he is doing, of course, is increasing the angle of attack still further by pulling the stick back, in essence slowing the aircraft still more with the speed control—the elevator.

What *should* he have done?

First, to recover from a stall close to the ground, it may not be possible or advisable to reduce the angle of attack sharply by shoving the stick forward, as this will cause a dangerous loss of the remaining altitude while correcting the airspeed (reducing the stall angle). What must be done is to apply a *coordinated correction* with stick and throttle together. Judiciously done, you thus bring the airspeed back up where it belongs by reducing the angle of attack with the elevator controls, *at the same time applying power* to maintain altitude control.

Above all, the wing must be kept flying to maintain control and avoid stalling or perhaps spinning out of control altogether. This may require a shallow dive, or, if brought under control early enough, it can be accomplished with no altitude loss at all.

Later on we will discuss stalls further, showing what effect wing loading, temperature, and other factors have in their cause and in recovery techniques. The point is that there are many, many things about flying that enter into good airmanship—things that you must experience to fully understand.

This is why it is a good idea to attend ground school regularly and take advantage of a certified ground instructor's ability to drill you in the fundamentals of flight. Once you understand the basic aerodynamics of flying, there is no mystery when something happens in the sky, such as encountering a stall. You know immediately what has gone wrong and your reaction is instinctively correct— when you understand the *why*.

Taming the Stall

It is man's nature to abhor a fall; when the nose of an airplane begins falling, an untrained pilot will instinctively try to pull it back up—exactly the *wrong* thing to do. An early pilot named Lincoln Beachey (Fig. 3-1) was the first man to explore this phenomenon, after a number of rival exhibition pilots were killed in stall/spin crashes. Beachey took his frail little biplane up above the clouds, where there was plenty of room, and tried to duplicate the maneuvers that had killed his friends.

The first time he tried it, Beachey reacted as the others had, hauling back hard on the elevator controls when his pusher plane nosed down into an uncontrollable dive. Things began to happen fast; his cap blew off and flew back into the propeller behind him, the wind in the wires screamed at him to let go, and the ground below began rushing up at him in a spinning blur.

He knew then that this was the killer stall/spin that had claimed his pals, but he didn't know yet what to do about it. In desperation, the brave little pilot (he was only five feet tall) grimly set his jaw and shoved the control wheel forward to see what would happen. As if by magic, the ship came out of the spin in a steep dive from which he was able to recover. Later on he tried it again and again, perfecting his recovery technique without really understanding what was happening. Not until he was back on the ground

Fig. 3-1. Early birdman Lincoln Beachey was first to explore proper method of recovering from a stall. (courtesy E.D. Weeks)

and discussed the recovery with others did he realize that he was reducing the angle of attack; he was eliminating the wing stall by going against instinct and shoving forward on the wheel instead of pulling back.

After that, pilots began tying pieces of string onto a front strut to show the direction of the relative wind. About the same time, the Wright Brothers invented a crude angle of attack indicator, a simple weather vane that always pointed into the slipstream. A scale behind it measured the angle of attack (Fig. 3-2).

The idea was forgotten for years when pilots began relying on airspeed indicators as stall-warning devices. It seemed logical that if a plane stalled at a certain airspeed, all you had to do was keep above that speed to avoid a stall. They were decidedly wrong!

Subsequent investigators showed that a wing can stall at virtually *any* airspeed. The determining factor is not speed, but angle-of-attack. For example, an aerobatic pilot who wants to do a snap roll first dives a little to increase his airspeed, gathering enough energy to make the ship do a complete horizontal high-speed spin. Then he yanks back hard on the stick, throwing the wing at the relative wind at a high angle, and at the same time kicks hard on either left or right rudder. The result is that if he kicks with his left foot, he swings the nose of the ship left at the same time he is stalling it so that the left wing stalls completely. The right wing, moving forward in a helical path, maintains airspeed *above* the stall. With the lift on the right wing pulling up, the ship snaps through a corkscrew path, a snap roll resulting from a partial high-speed stall.

Today the FAA is exploring several new types of angle of attack indicators to decide if they should be installed as "primary" instruments to supplement airspeed indicators. All work on the same principle of measuring the angle at which the wing meets the air (the direction of flight) with different kinds of angle sensors.

Right after World War I, the Sperry Gyroscope Company manufactured a primitive stall-warning indicator in which an electrical contact closed at a preselected airspeed, causing a warning horn to blow. This type device is still in use today, but serves primarily to warn of low speed, not high angle of attack.

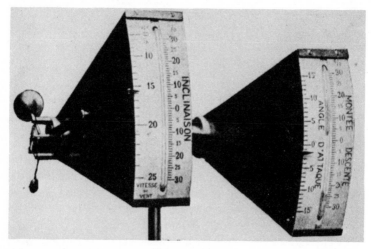

Fig. 3-2. Early angle-of-attack indicator of 1909.

Fig. 3-3. Modern-day angle of attack indicator is a new concept.

In the earlier days of flying, instructors taught their students to sense approaching stalls (and so higher angles of attack) by noting a mushiness of the controls as the slipstream slowed down, a change in the pitch of the wind blowing over the struts, and the visual position of the nose against the horizon, which is the pilot's basic flight reference line.

All these things help, of course, but seat-of-the-pants flying is not enough, particularly today when small aircraft have twice the horsepower, wing loading, and speed of earlier trainers. Nor is airspeed enough, for there is a definite "lag" in the airspeed indicator reading; it is in effect a "history" instrument that tells you what your airspeed was a moment ago. This happens because the airspeed indicator is essentially a pressure instrument, and it takes time for pressure changes to be reflected in movement of the needle over the dial.

Conversely, angle of attack indicators are "real time" instruments that respond so fast to changes in angle of attack that manufacturers build dampening units into their circuitry so that they will not confuse the pilot by flicking back and forth in turbulent air.

Under study by the FAA are three types of angle of attack indicator presentations. In one, a needle sweeps around a dial (Fig. 3-3); in another a light moves up and down a vertical scale as the angle changes; in a third a pointer moves from right to left as the angle increases.

Unlocking the Secret of Flight

A discussion of instrument flying will come later, but it is well in the beginning to understand how and why an airplane flies before starting actual flight training. In the early days of invention, flying machine builders attempted to gain lift by building simple flat-plate "airfoils" in which air pressure beneath the inclined planes pushed them upward. Let's look back for a moment to try to understand why it took so long for scientists to grasp the basics of heavier-than-air flight. By doing so, we can more readily grasp the real reasons why an airplane flies—how, in essence, it stays up in defiance of gravity.

An Englishman, Francis Herbert Wenham, founder of the Aeronautics Society of Great Britain, was a brilliant investigator who built the world's first wind tunnel in 1871 in an effort to promulgate aerodynamic laws on which a true science of aeronautics could be founded. A simple trunk 18 inches square and 40 inches long, Wenham's tunnel contained a fan (powered with a steam engine) producing a 40-mph wind.

Unaware of the need for curved wing surfaces, Wenham blindly went ahead experimenting with flat-plate surfaces set in the tunnel at angles from 15 to 60 degrees. If nothing else, Wenham did find out what *not* to do. Unpredictable turbulence upset his anticipated tables of pressure data, yet he did discover one highly important fact: At smaller angles of attack, the force of lift exceeded the force of drag to a far greater extent than he had believed possible.

The first to understand the real value of the curved wing was another Britisher, Horatio Phillips, who built his own wind tunnel and in it made the historic discovery that an area of low pressure forms above a cambered airfoil when air flows over it. Phillips' tunnel, far more sophisticated than Wenham's, incorporated a unique venturi tube design in which steam, sprayed through fine nozzles under 70 pounds of pressure, created a suction that produced a headwind of 60 feet per second (Fig. 3-4).

Phillips developed and patented a number of wing curves, some featuring the downward-swept leading edges that would become popular with Louis Bleriot and other pioneer designers of aircraft shortly after the Wright Brothers first flew in 1903. One of his wing sections actually attained a ratio of lift over drag (L/D ratio) of 10:1, unequalled for many years.

Once it had become clear that lift was produced essentially by a partial vacuum on top of the wing and not from positive pressure beneath it, other researchers began studying this phenomenon to find out why. A Danish scientist, H. C. Vogt, had noticed this effect in studying the beautifully curved sails of racing yachts; the sails apparently obtained much of their forward thrust from low-pressure areas on their leeward side, away from the wind. It was clear that the same laws of aerodynamics applied to bird wings and yacht sails.

More recently, aerodynamicists discovered that another phenomenon, the vortex flow around the wingtip (from the positive pressured area below to the negative pressure area above) is a major factor in the dynamics of flight. These tip vortices are actually responsible for the major portion of drag produced by a wing moving through the air, or *induced* drag. Such vortices trailing behind fast jets are a hidden danger to small aircraft; they form whirlpools of air that can flip a trainer upside down. (Vortices also can be harnessed, as in the design of special delta-shaped wings that are virtually stallproof—an inherent part of the design of modern supersonic aircraft such as the Concorde SST.)

Once it was established that a cambered wing produced the most lift and least drag, researchers like Otto Lilienthal began compiling statistical data to find out which curve was best. But not until the Wright Brothers began a systematic investigation with homemade wind tunnels and devices mounted on a bicycle were discrepencies in established "laws" rooted out and a new science of aerodynamics formulated.

For example, a key piece of the aerodynamic formula puzzle, the standard coefficient of pressure (k), had been established at 0.005 in 1759 by John Smeaton, an Englishman, who worked it out studying windmills. Nobody ever challenged it—that is, not until Orville and Wilbur Wright mounted a portable pressure test rig on a bike and rode it through the streets of Dayton, Ohio, checking and double-checking Smeaton. The real value for k, they discovered, was nearer 0.0033. With this value established (they were within

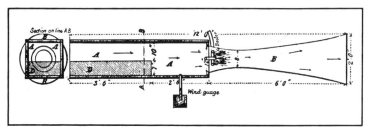

Fig. 3-4. Horatio Phillips' 1884 wind tunnel was six feet long.

Fig. 3-5.Wright Brothers' big achievement was teaching themselves how to fly. Orville here solos December 17, 1903.

6/10,000ths of being absolutely right!) they plunged ahead in 1901 and, using their own wind tunnel, tested some 200 different wing shapes until they found the one that gave the highest L/D. The one they chose could lift 764.78 pounds with a drag of only 80.34 pounds in a 30-mph wind, with the wing at a 5-degree angle of attack. This wing, with an L/D ratio of almost 10, was used in their famous Number 3 glider in which they made nearly a thousand successful glides down the Kill Devil Hills in the fall of 1902.

In making those glides, the Wright Brothers did more than prove their aerodynamic theories—they taught themselves to fly. Where other inventors like Octave Chanute and Samuel Langley had come close to success, they had failed basically because they did not recognize that flying is an *art*, a thing to be learned progressively (Fig. 3-5).

Today, however, although the Wright Brothers have given us a basic understanding of aerodynamics (including the three-torque control system of elevators, rudder and ailerons, or wing-warping), we too often see student pilots taking flying lessons with no real appreciation for the beauty and integrity of heavier-than-air flight.

Ground schooling in the simple laws of aerodynamics is essential to becoming a first-rate pilot. Without such a study, you cannot appreciate just what holds your airplane up, and how it is able to fly upside down or on edge in aerobatics. If you could look out over your wing and see *how* the air flows and *where* the stall begins *when* the angle of attack starts to get too steep, you would sense the vise-like grip of wind over wings that makes airplane flight safe and sure.

In fact, there *is* a way to visualize aerodynamics in action, a system used not only by wind tunnel investigators but by dedicated flight instructors and test pilots who want to see what is happening as an airplane flies through various maneuvers. Tufts of wool yarn, one or two inches in length, are attached to the upper surface of a wing in row after row (Fig. 3-6). A simple way to do this is to get a ball of yarn and wrap it around the wing, front to back, spacing the spirals an inch or two apart, from the wingtip to the wingroot. Then strips of masking tape are stuck over the yarn, parallel to the leading edge, at intervals of three to four inches. The yarn is then cut with scissors just ahead of each strip of tape. The result is a "fur" of wool tufts that are free to flow backward straight and smooth under the airflow when the wing is in normal flight.

Once the pilot brings the control stick back and begins to raise the nose of the ship, increasing the angle of attack of the relative wind blowing over the wing, you will notice the beginnings of a stall.

In most wing designs, the stall commences near the trailing edge of the wingroot, progressing steadily forward and outward as

Fig. 3-6. Homebuilder Don Janson taped wool tufts all over his Smith Miniplane racer to look for any unwanted turbulence.

Fig. 3-7. Root stall is evident in this photo of a tufted wing reaching critical angle of attack.

the angle of attack increases. In modern wing design, the outward section of the leading edge is lower than the inboard portion so that it flies at a lower angle of attack. Thus you will actually see that air continues to flow smoothly over the outer wing even while the inboard part of the wing is almost completely stalled as evidenced by a wild and frantic dancing of the wool tufts (Fig. 3-7).

There is a definite reason for this—it is desirable that the ailerons remain operative as long as possible, even in a stall landing, to give the pilot good lateral control. A laminar flow of the wind over the wingtips and ailerons provides this control even though the rest of the wing has stalled sufficiently to permit the aircraft to land.

This is also the reason why the FAA now encourages pilots to use aileron and rudder in a coordinated manner in recovering from stalls. Some pilots argue that the use of ailerons can be dangerous in recovery from a stall—and in some cases they are right. In one much-used World War II trainer, the Ryan PT-22, the whole wing stalls at about the same time; here, use of aileron in stall recovery is dangerous (Fig. 3-8).

This is so because of the following sequence of events: As the PT-22 wing stalls and the nose of the aircraft starts to drop, any minute unevenness in the stalled condition of the wing can cause

Fig. 3-8. Use of aileron in stall recovery of PT-22 trainer can be dangerous to your health.

one wing to drop ahead of the other so that a roll is induced around the horizontal axis—the start of a spin. In PT-22 stall recoveries, instructors warned their cadets *never* to use aileron to "pick up" the low wing, because to do so only further increases the angle of attack of that stalled wing by depressing the aileron control below the wing's chordline. Instead, standard recovery procedure was to hold the nose straight with the rudder alone and reduce the angle of attack with a sharp forward movement of the control stick, depressing the elevator surfaces into the slipstream and putting the ship into a corrective dive.

Many times I have watched a look of surprise and amazement spread over the face of a student pilot who found himself snapping inverted unexpectedly simply because he used aileron pressure to bring up a low wing in a stall. Even today, with modern wing designs that provide aileron control through the full range of a wing stall, hard misuse of aileron control can break the laminar flow to the point where the same thing occurs—a vicious snapdown of the stalled wing.

It is because of such phenomena that students need good flight instructors to show them what is really happening in flight. Only with this understanding do you get the most benefit from flight training. You may go ahead and get your private (or even commercial) pilot's license and fly for years without getting into trouble, but unless you fully understand what is happening, you are simply riding your luck.

Today, the FAA still recognizes the inadvertent stall/spin as one of the major causes of fatal crashes—the more so in an age of aircraft with higher wing loadings, higher stall speeds, and more horsepower. This is the reason why in recent years there has been a renewed interest in aerobatic instruction; pilots themselves sense a real need to know the full limits of the planes they fly in order to be able to do the right thing when trouble occurs.

Let's take a typical stall/spin accident report in which a light-plane crashes on takeoff. The pilot is flying out of a short field at high elevation, where the air pressure is below that which he is used to. Because of the lower air pressure, his wings do not get the same lift energy from the thinner air as when taking off at sea level at the same ground speed, and so his takeoff speed must inevitably be higher.

He rotates (pulls back on the stick to lift the nose and place the wing at a higher angle of attack) but the ship doesn't begin to climb as it usually does at a lower altitude airport. Instead, it mushes ahead. The pilot staggers off the ground under full engine power, his wing barely flying. Dead ahead of him looms a line of trees. It's too late now to get over them, so he tries to turn and miss them. His right wing, already virtually riding the edge of a stall, is further stalled when the pilot shoves the stick hard left to start his turn away from the trees. Instead of turning to the left, the now-completely stalled right wing drops and he is in a deadly spin.

Of course, there were other contributory factors involved in this accident. The pilot used poor judgment in attempting a takeoff from a high-altitude short field in the heat of the day; he might have waited until late afternoon for the air to cool. He rotated too soon, before the airspeed was sufficient to produce adequate lift. He tried to force his ship into the best angle-of-climb before it was at the proper airspeed. And lastly, when there was nothing left to do, he could have flown on into the trees and possibly survived a lesser crash than spinning into the ground. If he were sufficiently skilled, he might even have made a gentle turn without misusing the aileron and so destroying lift completely.

You can understand now why it is that a pilot needs a thorough course of flight instruction *and* ground instruction to become aware of the limitations of his aircraft and himself. In modern general aviation aircraft, you have to work hard to get into trouble—but you still can! Once you know what you are doing and operate your ship within the limits of your own ability, you are safer than you are driving a car on a crowded freeway. The sky is yours to enjoy—if you understand and conform to the Federal Aviation Regulations and to the simple laws of aerodynamics that your ground and flight instructors will explain to you.

There are no mysteries about flying. As we said earlier, it doesn't take a Superman to fly. Once you master the basics of how and why an airplane flies, you are on your way to a glorious adventure in flight.

Chapter 4
Your First Flight

Once you have decided on the right airplane to buy, selected a good flight instructor, and signed up for a ground school course of study in theory of flight, meteorology, navigation, and Federal Aviation Regulations, you are ready to begin the real fun of flying—handling an airplane yourself.

There was a time when flight instructors considered themselves very dashing fellows—to be looked up to and feared. They strode about the flight line, leather helmet earflaps turned upward, goggles sitting above their brows, a cigarette dangling from thin, hard lips. They leaned forward when they walked, as if in the teeth of a gale, largely from the habit of balancing a seat-pack parachute bobbing on their backsides.

With a wild gleam they lured you into the cockpit, flipped away the cigarette, and proceeded to show you what hot pilots they were. You sat in the rear cockpit, clinging to the sides of the ship for dear life, sure this character was out to kill you—or at least shame you with his superior skill and your insignificance.

At a mile above the green-and-brown patterns of fields and roads and fences, you peeked over the side, then quickly shut your eyes. When you opened them, you saw your instructor's face leering at you in the rear-view mirror. He picked up the gosport speaking tube, a sort of stethoscope with a funnel in one end and your ears in the other, and screamed:

"Hang on, sonny boy! Check your safety belt!"

You knew then it was all a mistake. You never should have decided to take up flying. Better to have gone in for knitting or pogo sticks. The sky is no place for you, only for birds and tigers like the guy in the front cockpit.

Suddenly the horizon spun crazily and you felt a wild panic. Your instructor was putting you through the wringer, to "see if you can take it." He looped, rolled, and spun all over the sky, and only came down when you threw up. He leapt lightly to the ramp.

"When you clean up the ship and yourself, come on into the ready room," he snapped. "I want to have a few words with you!"

You staggered off, climbed into your car, and drove home to take a shower, the idea of flying forgotten. You felt lucky to be alive.

The New Breed of CFI

Fortunately, those days are gone now, and the art of flight instruction demands only men and women of skill, intelligence and understanding. The FAA has weeded out the idiot sky teachers who gave way to sadistic impulses and did more harm than good.

Not that all CFIs are tender-hearted characters, for their job is a tough one, demanding endless patience and a constant awareness of the needs of their students. Jack Frymeier, of Ohio State University's Department of Aviation, emphasizes at Flight Instructor Seminars that the best instructor deliberately diversifies his instruction technique to fit his students' needs—not his own. His approach may be supportive, directive, discussive, or persuasive—whatever works best with a particular student.

A fellow Ohio State mentor, Jack Eggspuehler, participated in a survey to determine why it was that 60 percent of new flight students never reached the level of pilot certification. In the year this survey was conducted, 128,000 student pilot licenses had been issued, so statistically 76,800 potentially active pilots and plane owners simply gave up.

Everybody has a different answer to this problem—the National Aviation Trade Association, the FAA, fixed base school operators, instructors—everybody blamed everybody else. So Eggspuehler and his associates simply visited a large number of flight schools and selected a list of names of would-be pilots whose old logbooks were left gathering dust. Of 1052 ex-students approached, 28.4 percent replied to the questionnaire, a good return.

What they found did much to correct a problem of "pilot dropouts" that plagued the aviation industry for a long time. Once the causes were known, corrective measures could be taken. Many of the dropouts were lured back to complete their flight training, and were happy to do so.

The number one problem, it turned out, was simple economics. Half had spent more than $15 a week on flying as an extracurricular activity. Of these, 45 percent actually felt that the rates were reasonable, while another 40 percent admitted they would return to flight training if the rates were lower. This was highly significant to the aircraft makers, particularly when figures showed that 60 percent had initially thought of buying an airplane. This figure dropped alarmingly *after* solo, when student pilots became aware of unexpected costs. At that point, only 8 percent wanted to own airplanes.

The second reason for dropping out was *time*; while 57 percent flew regularly every week, 75 percent found they were spending from two to three hours a week travel time getting to and from the airport. Instructors came in for heavy criticism, in wasting their students' time by simply not being on time themselves for lessons.

Paperwork was the third problem that appeared in the survey, and 28 percent admitted that preparation for FAA examinations was the most difficult part of flying. "Most people are afraid of failure," said Eggspuehler. "The FAA inspector represents a threat."

Incidentally, FAA inspectors themselves are the first to admit to this, but they consider "inspectoritis" a good thing. "We want a student to be tense on a check ride," one told the author. "Otherwise, we won't know how he will act in an emergency, under pressure. Of course we make allowances for this, because intimidation is a very real thing." He remembered one good flight student who paled when he simply walked up and said, "Hello!"

"I could see his brain working," he smiled. "The poor guy was thinking, 'My God, what did he mean by *that*?' "

The fourth reason for student pilot dropouts, Eggspuehler found, was what he called the "Mama" influence. Of the subjects interviewed, 63 percent were married, and of these, 35 percent felt flying was either too expensive or unsafe. "One interviewee replied: 'We both fly. My wife flies a broom!' "

The fifth problem was worry over the elements, with 41 percent fearful over high winds and 59 percent admitting they would not fly in poor visibility.

The sixth trouble was with the instructors themselves. Many didn't keep appointments, were impatient and lacked understanding, or were simply deficient in their ability to meet emergencies.

This last point came as a shocker, and has resulted in closer

FAA supervision over CFIs, who now are required to have their licenses revalidated every two years.

The last three problems faced by students pilots were dislike of exposure to emergency procedure instruction, frustration in accomplishing difficult routine flight maneuvers such as crosswind landings, and unavailability of airports close to home.

The conclusion of this interesting survey was that student pilots seemed to have gone blindly into flight training with the idea that it was all just recreation and no hard work. The corrective first step, Eggspuehler told one CFI seminar I attended, must be taken by the instructors themselves: Stop being intense and help the student through the difficult early part of his orientation in the sky.

"A student who bounces a landing hard may get a feeling of failure unless his instructor helps him know what he did wrong and correct himself," he explained. The student-instructor relationship indeed is an intimate one. Sarcasm and bad humor have no place in the cockpit of a trainer.

Today you will find most instructors conforming to a changed philosophy and adhering to the newly adopted Code of Ethics of the National Association of Flight Instructors. Once you have found a competent CFI to work with you, there is no reason why you should not go right ahead on schedule, complete the required hours of instruction, and emerge to join the other hundreds of thousands of pilots who make the sky their home.

Fig. 4-1. Skillful CFI doesn't try to scare his student. Instead, he establishes rapport, using hand-talk to demonstrate.

Now let's talk about that first flight, for the benefit of those readers whose only experience in flying has been seated comfortably in a big jetliner with a cold martini and a first-run movie.

The first thing your instructor will do is chat with you a while to establish a rapport and put you at ease on a friendly one-to-one basis (Fig. 4-1). He'll size you up just as you size him up, each taking the other's measure. This is the guy to whom you are entrusting your life, now and in the future, because how well he drills you with flight fundamentals will be reflected in how you perform in an emergency, when you're up in the sky alone.

Perhaps a brief chalk-talk at a blackboard in the flight office will serve to brief you on the airport traffic pattern you'll be flying (Fig. 4-2). When that is over, you'll go outside for a "walk-around" inspection of your airplane to make sure it's airworthy.

Preflight Inspection

Perhaps you'll start at a wingtip, checking it for any visible damage. Then you'll look over the aileron, inspecting the hinges and cable connections (or pushrods, as the case may be) for signs of unusual wear or looseness. If there are holes in the wing (inspection plates may be missing), look inside for bird nests. Birds love to raise their own fledglings inside a comfy wing, either fabric or metal skinned. (I onced had my Luscombe grounded a full month one spring until a sparrow family decided to fly off.)

Fig. 4-2. Flight Instructor Don Prosser uses blackboard chalk-talk to emphasize airport traffic pattern flying.

Fig. 4-3. Preflight inspection includes check on tailwheel.

Do not think of a preflight line inspection as a ritual to go through in a hurry just for appearance's sake. Here is where hidden troubles can be spotted in time to take corrective action before becoming airborne. More than one pilot has made an emergency landing because he forgot to remove the gust-locks from his control surfaces at this point!

Your inspection should include an examination of the underbelly of the fuselage to look for traces of oil leaks or battery acid corrosion. Look underneath the empennage (tail surfaces) to see whether rocks have torn holes in fabric elevators while taxiing.

Test the tailwheel springs and the swiveling gear, to make sure they are free from trouble (Fig. 4-3). Complete the walkaround with a visual inspection of the rudder surface, the other side of the fuselage, and the opposite wing. If you are planning a night flight, be certain to check all navigation, beacon, and landing lights.

Look your propeller over for nicks, bird droppings, or lamination separation in wooden blades caused by moisture. Lift the engine cowl and check underneath for bird nests or other foreign material that might constitute a fire hazard. Are the electrical wires in good shape? The exhaust manifolds?

Check your crankcase oil level, and be sure the filler cap is on tight after checking it (Fig. 4-4). Open your main fuel valves and check the drain plug underneath the sump bowl for water that may condense within a partially full wing tank and run down the feed lines. It is good practice, incidentally, to always fill the fuel tanks after a flight to prevent moisture from forming inside the tanks on humid days.

There are other important things to look for on your line inspection. Don't rely on fuel gauges; they can be wrong (Fig. 4-5). Remove the fuel tank caps and make a visual inspection. Examine the pitot tube opening, and be sure the dust cover is removed. See that windows are clean, and all cowl access doors are secure. In cold weather, pull the propeller through by hand (first making sure switches are off!) for two or three revolutions to check oil viscosity, and cylinder compression (Fig. 4-6).

Once you are satisfied that the exterior of your aircraft is in good shape (Fig. 4-7), that the tiedown lines have been cast off, and that the landing gear oleo struts are properly inflated or filled with hydraulic fluid, turn your attention to a thorough cockpit preflight check.

This will vary among different types of aircraft, but by follow-

Fig. 4-4. Pilot checks oil level with dipstick in Cessna 170.

Fig. 4-5. Don't believe the fuel gauge. Instead, remove the filler cap and visually inspect avgas level.

ing your manufacturer's recommendations, you will cover the main points to make sure you are ready to fly. Most pilots adopt some kind of a mneumonic (memory) system of code letters to remind them of the key things to check over. One such is TMPFF—Trim, Mixture, Propeller Pitch, Fuel, and Flaps.

Your instructor will tell you what to look for in your airplane, and a standard preflight cockpit check will cover the full instrument panel, the electrical system, radio equipment, and flight controls. Make sure the latter are free from obstructions by moving them *fully* in *all* directions. See that the trim tab is set in takeoff position, mixture control full rich, pitch control in fine position (if your propeller is adjustable, fuel selector valve on the proper (fullest) tank, and flaps positioned where you want them for takeoff.

Startup and Takeoff

Seated, safety belt and shoulder harness fastened (Fig. 4-8), you will now go through the procedure of starting engine—brakes set, engine primed (if necessary), throttle cracked (slightly opened), magneto switch on. For safety's sake, open a window and call out. "Clear!" as a warning to anyone in front of the plane that the propeller is about to start turning. Then turn on the master and starter switches.

Once your engine starts running, check to see that the engine

Fig. 4-6. Pilot Nick Stasinos demonstrates wrong way to handle a prop, with fingers hooked around blade. A backfire could hurt!

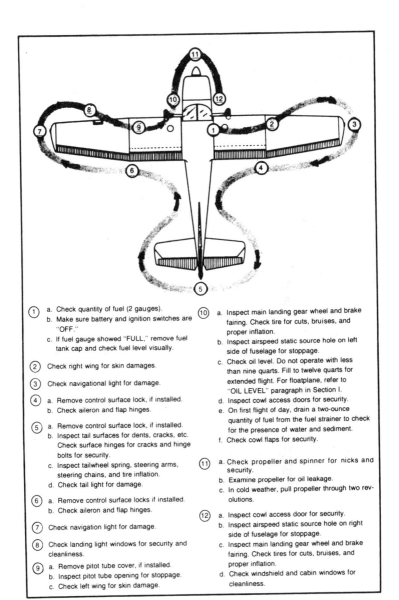

① a. Check quantity of fuel (2 gauges).
   b. Make sure battery and ignition switches are "OFF."
   c. If fuel gauge showed "FULL," remove fuel tank cap and check fuel level visually.

② Check right wing for skin damages.

③ Check navigational light for damage.

④ a. Remove control surface lock, if installed.
   b. Check aileron and flap hinges.

⑤ a. Remove control surface lock, if installed.
   b. Inspect tail surfaces for dents, cracks, etc. Check surface hinges for cracks and hinge bolts for security.
   c. Inspect tailwheel spring, steering arms, steering chains, and tire inflation.
   d. Check tail light for damage.

⑥ a. Remove control surface locks if installed.
   b. Check aileron and flap hinges.

⑦ Check navigation light for damage.

⑧ Check landing light windows for security and cleanliness.

⑨ a. Remove pitot tube cover, if installed.
   b. Inspect pitot tube opening for stoppage.
   c. Check left wing for skin damage.

⑩ a. Inspect main landing gear wheel and brake fairing. Check tire for cuts, bruises, and proper inflation.
   b. Inspect airspeed static source hole on left side of fuselage for stoppage.
   c. Check oil level. Do not operate with less than nine quarts. Fill to twelve quarts for extended flight. For floatplane, refer to "OIL LEVEL" paragraph in Section I.
   d. Inspect cowl access doors for security.
   e. On first flight of day, drain a two-ounce quantity of fuel from the fuel strainer to check for the presence of water and sediment.
   f. Check cowl flaps for security.

⑪ a. Check propeller and spinner for nicks and security.
   b. Examine propeller for oil leakage.
   c. In cold weather, pull propeller through two revolutions.

⑫ a. Inspect cowl access door for security.
   b. Inspect airspeed static source hole on right side of fuselage for stoppage.
   c. Inspect main landing gear wheel and brake fairing. Check tires for cuts, bruises, and proper inflation.
   d. Check windshield and cabin windows for cleanliness.

Fig. 4-7. Line inspection covers complete aircraft exterior. (courtesy FAA)

gauges are operative. Oil pressure should rise; if the weather is cold, let the engine idle until the oil temperature begins to show an increase.

During this time, if you are flying from a controlled field, run a radio check of the frequencies you intend to use—ground control, tower, and departure control. By now you are ready to go. In a later chapter on communications we will go into proper radio procedure, but suffice it to say here that you must get approval from ground control to taxi into takeoff position at the downwind end of the active runway. You will receive such information on field conditions as

Fig. 4-8. Dorothy Joslyn feels secure with shoulder harness fastened in the family's Bellanca.

necessary—wind direction and strength, field barometric pressure (altimeter setting), and warnings of any existing hazards, such as construction equipment parked near a taxiway.

With your instructor beside you, directing your actions and telling you what to expect next, you advance the throttle and with fresh confidence feel the ship begin to roll. You guide it with your rudder pedals, and, if necessary, you keep the nose moving gently left and right so that you can always see ahead.

It may be that you will be taxiing in a strong crosswind, a condition in which there is a possibility of a strong gust getting under a wing and upsetting your ship (Fig. 4-9). It is best to hold the aileron control *into the wind*, from whichever quarter it is blowing, to present the least underneath surface to gusting. Similarly, taxiing downwind, remember that a gust under your tail could put you on your nose, so hold the stick forward (or back, taxiing into the wind, to hold the tail down).

There may be need to use your brakes in a crosswind to keep your ship from "weathercocking" into the wind, but try to use the brakes as little as possible to save wear and tear.

You are now in position at the runup area, next to the downwind end of the active runway, and it is time for a last preflight cockpit check. Once more—TMPFF! or whatever system you use. Also check your carburetor heat, noting that a slight drop in rpms occurs when it is applied. Run up your engine to around 1500 rpm and check your left and right magnetos separately. If one bank drops in rpm more than 100, it may indicate a fouled spark plug. A maximum power runup finally tells you that you have sufficient power for takeoff (the rpm should exceed 2000 in most lightplane engines).

Now call the tower for takeoff clearance, and don't forget to take a good look for incoming traffic, which has the right of way over planes taking off. The tower operator is safe inside his tower; it's *your* neck if someone lands on top of you! You are the captain of your ship, legally responsible for its safe operation.

When cleared for takeoff, taxi smoothly onto the runway, line up with some distant object on the horizon (to give your eyes infinity focus), and advance the throttle evenly to takeoff power. If the wind is down the runway, there will be no trouble holding the aircraft straight with quick and gentle rudder pressures, offsetting both inadvertent swerves and torque forces from the engine and propeller when your tail begins to lift.

Normally, the aircraft will assume its own takeoff attitude under the force of slipstream pressures until the speed of the wind

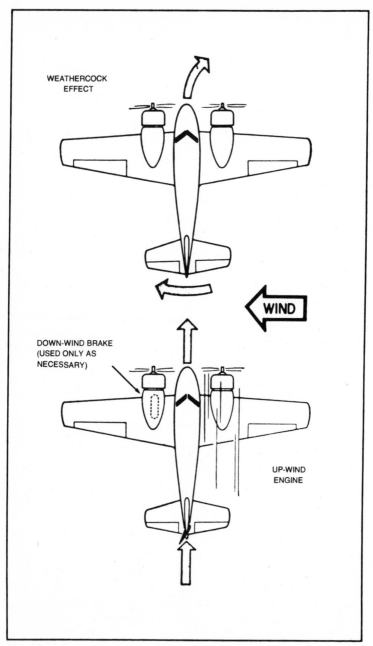

Fig. 4-9. Taxiing twin-engine aircraft in strong crosswind requires extra power on upwind engine, opposite rudder against weathercocking tendency. (courtesy FAA)

over the wings produces sufficient lift to get you airborne. In crosswind takeoffs, and in some heavier aircraft, you may want to hold the ship on the ground longer to get your airspeed well above stalling speed before letting it fly, thus avoiding a sideways skittering. At other times, you will want to lift off at minimum flying speed (as in taking off from a sandy or soggy field) to eliminate ground drag.

Climbing Out

However you do it, you will come to a velocity that pilots call by different names—V_1 (critical engine failure speed), V_2 (emergency takeoff climb speed), V_r (rotation speed, V_{mc} (minimum control speed) or V_u (unstick speed). You are now making like a bird!

Climbing out of the airport traffic pattern, you have other choices such as V_x (best *angle* of climb) and V_y (best *rate* of climb). You would use the former airspeed to make a steep climb over an obstruction, the latter to climb to cruise altitude most efficiently (Fig. 4-10).

Let's forget aerodynamics and velocities and pattern flying and radio procedures for a moment and look around at the transforma-

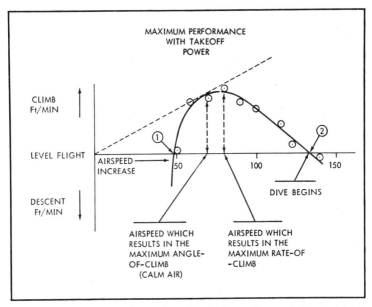

Fig. 4-10. Graph shows airspeed use to obtain maximum climb with takeoff power. (courtesy FAA)

tion the Earth is undergoing. You are entering the airman's world, where cities and towns and distant mountains and rivers blend into breathtaking panoramas.

Unfortunately, in our crowded skies today we cannot simply go sightseeing or joyriding through this wonderland oblivious of others. We all share the same sky in a very democratic fashion. As we have said, the 250,000 miles of Federal Airways are there for big and little planes alike. Unless you are following a Victor Airway on a cross-country flight, either under IFR or VFR regulations, you will stay clear of these airways, which are shown on aeronautical charts, and go find a local practice area to feel out your new airplane.

Here is where you begin to hear the song of the sky, where you will win your wings as you learn more and more about how to make your aircraft respond to gentle pressures. Here is where you will become familiar with pilotage and map reading, and learn to relate to the ground below you. Precise navigation and use of your radio will come later. Now you want to get the feel of flying and taste the glorious freedom of being in positive control, making your craft do what you want.

Your instructor is along to help you and guide you, but *you* are doing the flying! You're on your way!

Chapter 5
Fundamentals of Flight

The FAA lists 26 procedures and flight maneuvers with which you should develop competence prior to taking your check ride for a private pilot license. Here they are:

1. Airplane Familiarization.
2. Preflight Operations.
3. Familiarization Flight.
4. Use of Radio for Communications.
5. Straight-and-Level Flight, Turns, Climbs, and Descents.
6. Slow Flight and Stalls.
7. Coordination Exercises.
8. Ground Reference Maneuvers; Traffic Patterns.
9. Takeoff and Departure Procedures.
10. Approach and Landing Procedures.
11. Stalls from Critical Flight Situations.
12. Steep Turns.
13. Crosswind Landings.
14. Short- and Soft-Field Takeoffs and Landings, Maximum Climbs.
15. Power Approaches to Full Stall Landings, and Wheel Landings.
16. Turns to Headings and Recovery from Unusual Attitudes (by instrument reference only).
17. Emergencies.
18. The Solo Flight.

19. Practice Area Familiarization.
20. Cross-Country Flight Planning.
21. Pilotage; Map Reading.
22. Dead Reckoning; Use of the Compass.
23. Use of Radio Aids for VFR Navigation.
24. Obtaining Emergency Assistance by Radio.
25. Unfamiliar Airport Procedures.
26. Plotting Alternate Courses in Flight.

In addition, applicants for a commercial flight certificate should attain competence in the following:

1. Eights on Pylons.
2. 720° Power Turns.
3. Steep Spirals.
4. 180° Accuracy Landings.
5. Chandelles.
6. Lazy Eights.
7. Night Flying.
8. Instrument Flight Instruction.
9. Spins. (You do not have to demonstrate spin recovery for a private or commercial pilot certificate, unless you are qualifying for a flight instructor rating.)

While we have touched on a few of these procedures and maneuvers in earlier chapters, let's go through the list step by step and discuss what will be expected of you by your flight instructor and FAA examiner.

Airplane Familiarization

Your instructor will make sure that you know what all the main parts of your airplane are for and how they work. When you have completed your line inspection, he will see that you are comfortably seated in the cockpit (this is essential to good flying technique). Fastening your seat belt when you first enter the cockpit is good practice.

While at first the instrument panel may look like a confusing maze, don't worry. Your CFI will monitor the dials for you at first, directing your attention to them one at a time until you are completely familiar with them.

Placing your feet on the rudder pedals properly and holding the control wheel gently are essential. While they will feel sloppy when the aircraft is parked, once the airflow grips them firmly, they will feel different. You will learn then to fly by control *pressures*, not *movement* (except in some aerobatic maneuvers).

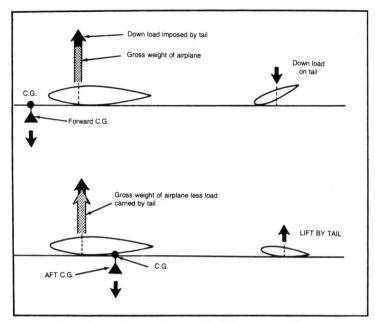

Fig. 5-1. Proper loading is essential for safe flight. (courtesy FAA)

It is well to understand that an airplane properly trimmed for straight-and-level flight will "fly itself" (Figs. 5-1, 5-2). If you displace the controls and let go, the airplane will return to its original attitude and heading. Some old-timers will confuse you with stories of how the controls "change function" in certain maneuvers—that in a vertical bank the rudder becomes the elevator, etc. *Don't believe it!* Think of control response in relation to yourself, not the horizon. (It is true that during inverted aerobatic flight there is a reversal of aileron function in turns, but only in relation to the horizon. Also, flying inverted, you make the nose of

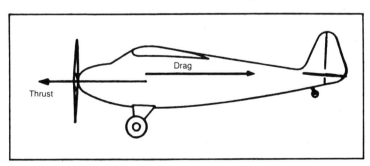

Fig. 5-2. Thrust and drag are equal in level flight. (courtesy FAA)

your ship go up by pushing *forward* on the control stick, but in relation to the pilot, nothing has changed!)

Operation of the throttle in coordination with the elevator control is another basic coordination factor fully as important as rudder-elevator coordination. Now is a good time to discuss both. As you read earlier, the elevator is the *speed* control; the throttle the *altitude* control. Both are used together, however, in what is called *attitude* flying. To get the best rate of climb from level flight, or the best angle of climb, you want to use not only a predetermined airspeed but a predetermined *attitude* of the nose in relation to the horizon, together with the proper power setting. (This varies with different aircraft.)

Typically, in a lightplane, you will raise the nose above the horizon and apply full power to climb—stick back, throttle forward, in coordination. To glide, the reverse is true—throttle back, stick forward to depress the nose below the horizon.

To change from a climb to level flight, some instructions say, it is well to climb perhaps 150 feet above your cruise altitude and dive back. In doing this, you translate the extra altitude into extra speed (say, from a 60-mph climb to an 80-mph cruise), at the same time resetting your throttle to cruise power.

The engine maker will recommend the best cruise power setting in terms of rpms; properly trimmed for level flight, this will produce a predetermined airspeed. You may want to burn less gas on a long trip, and so fly at a lower rpm setting, extending your range at a sacrifice of speed.

Always handle the throttle smoothly, for sudden applications of power can overload your engine and shorten its life—and maybe yours!

What about rudder-aileron coordination? First, let's understand that the rudder is not there to steer your craft left and right in level flight, like a motor boat. Except in such maneuvers as side-slips, it is used solely to compensate for an aerodynamic phenomenon called *aileron drag*.

In driving an automobile around a corner, you turn the steering wheel in the desired direction and hold it there until the turn is completed. In an airplane you do it differently—you turn the wheel (or move the stick) in the direction you want to turn, and once the bank you want is established you return the control wheel or stick to neutral.

Let's suppose you use no rudder pressure at all when you do this and see what happens: To turn left you move the controls left, thus depressing the right aileron into the slipstream, raising the right wing into a bank. Right away, the increased relative angle of attack produced by the depressed aileron creates more drag (aileron drag). In turn, this extra drag tugs the nose of your ship to the *right*, not left!

Here is where you use your rudder pressure, moving the rudder surface left so that the slipstream strikes the left side of the rudder just enough to offset the right tug of that aileron's drag. These forces neutralize each other, so you enter your turn with no uncoordinated side loads imposed.

Finding out just how much rudder pressure to apply is simply a matter of trial and error. Generally speaking, it is directly proportional to how much aileron deflection you use. For a quick, sharp entry into a turn, you will use more "stick and rudder" pressure than you normally would in a slower, more gentle entry. And of course, rolling out of a turn is about the same as entering one, where rudder-aileron coordination is concerned.

There is still another factor involved here in making a good turn without losing altitude. It has to do with the fact that you need more lift in a turn than in level flight to equalize the force of gravity. In level flight, lift acts at right angles to the wingspan, and in a bank—think!—it acts the same way!

By drawing a simple parallelogram of forces, you can see that the vector of lift acting *inward* in a banked turn shortens as the bank steepens. Hence, to keep it equal to the constant gravity force, you must increase the angle of attack with back pressure on the elevator control or else end up in a descending spiral.

There's another bothersome force that comes into the picture in a turn. This is the *overbanking* tendency caused by the simple fact that the outside wing in a turn travels farther (and so faster) than the inside wing, creating more lift (Fig. 5-3). It is a small force, but it can cause the aircraft to keep rolling about its longitudinal axis into an ever-steepening bank. To correct for this overbanking tendency, slight "opposite aileron" pressure is necessary.

Hence, when entering a left turn, you will move stick and rudder left until the desired bank is reached. Then you will remove pressure from the rudder and apply a combination of stick forces— back pressure to offset the gravity effect and a touch of opposite aileron to prevent overbanking. The stick comes back and to the right in an "L" pattern, if you think of it as movement instead of pressure.

Recovering from your turn, the process is reversed—right

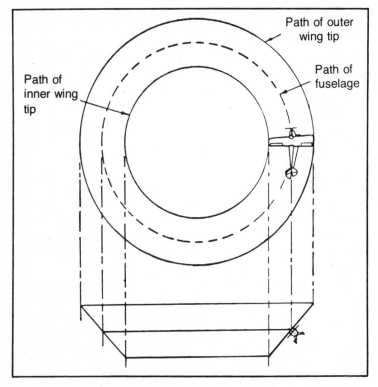

Fig. 5-3. A plane tends to overbank in steep turn because the outer wing flies farther and faster than the inner wing. (courtesy FAA)

rudder pressure, stick forward, and release right aileron pressure as you neutralize rudder. Simple? Maybe, but it takes practice, particularly in steeper banks.

If it occurs to you that we are spending too much time discussing the fine points of a simple thing like turning an airplane, remember this: All possible flight maneuvers are made up of only four basic maneuvers—climb, dive, turn and straight-and-level flight! Hence, once you have mastered these fundamentals, even the most complex aerobatics are made easier to execute.

Learning how to control an airplane properly and fly it smoothly and with good coordination, as important as it all is, simply prepares you to go on to the next step—flying the airplane in relation to the ground (as in airport pattern flying) or in relation to navigation aids on a cross-country flight. Here is where your knowledge of meteorology, navigation, communications, and other arts of flying comes into use.

Preflight Operations

In our earlier discussion we have touched on basic things to look for in preparing for a flight, so there is no need to explore this category further here, except to say that your instructor will brief you on items pertinent to your particular aircraft, on proper local taxi procedures, etc.

Familiarization Flight

This is the time when you and your instructor get to know each other and learn to communicate. It is a crucial time, one in which you will learn to enjoy flying or fear it, when you grasp the fundamentals of controls or are confused by them, when you locate and identify the reference points on the airplane that you will use henceforth. It's a good idea if you and your CFI make this a pleasant cross-country hop to a nearby airport for lunch. In this way you can ask questions about things new to you; on the way back you can increase your learning capacity by applying what he tells you.

I have found that early flight sessions are better spent if they are brief. Once the student gets the feel of the controls, experiments with them, and boots the ship around the sky to see what happens, apprehensions vanish. I remember one cadet at Falcon Field, during World War II, who sat frozen-faced in the rear seat of a Stearman, afraid to look left or right. Suddenly I was aware that my dual control stick was rigid. In alarm, I checked the locking system; it was free. I recognized the problem immediately—the poor student was squeezing his stick hard enough to bend steel.

I breathed a sigh of relief, but something had to be done fast, and I didn't want to add to his panic. Over the intercom, I asked him to hand me a map of the training area he had with him. His attention diverted, he relaxed a bit, and then I asked him to place the tip of one finger on top of the stick to see how easily the ship could fly by itself. Soon he was grinning widely. He had discovered that if he let go of the control stick, the airplane wouldn't fall down.

Such apprehensivess, incidentally, is the exception to the rule, and usually stems from some bad prior experience such as a "wring-out ride" with a sadistic instructor in the past. The average beginner finds his first flight an exhilarating experience, and enthusiasm at this point is a great help to meeting more complex assignments.

Use of Radio for Communications

Later on, in discussing radio navigation, we will go more

deeply into the whole story of radio communications, for their functions are closely related; on a cross-country flight, the first radio contact is with Ground Control for taxi instructions.

Many flight schools today like to introduce a student to radio communications at first in a flight simulator on the ground, so that he can devote his full attention to it. In actual flight, the student at first has his hands full getting started on aircraft familiarization. A lot of radio chatter with the tower can be so utterly distracting that he may blank out and remember nothing. For this reason, many instructors operating from radio-controlled airports prefer to do the radio talking for the student at first so he can concentrate on flying. Others prefer to start their students off right away on radio work, for today it has become an integral part of airmanship.

There are simple rules to follow: Talk clearly, slowly, and succinctly. Identify yourself by aircraft call number, state your message, and *shut up*. Too many pilots fail to talk efficiently, while others try too hard and become tense. If you talk easily over the telephone, there's no reason to be scared on an aircraft radio.

To operate an aircraft radio legally, a student pilot should first apply for a Restricted Radiotelephone Operators Permit from the Federal Communications Commission. While correct phraseology in radio transmissions is desirable, don't be afraid of saying it wrong. In fact, FAA traffic controllers will help you out if you say: "Hello Cupcake Tower, this is student pilot Jones in Cessna Zero Zero One. I wanna come down!"

Straight-and-Level Flight, Turns, Climbs, and Descents

At the outset, says the FAA, "it is impossible to emphasize too strongly the necessity for forming correct habits in flying straight and level."

We have discussed already how to trim your ship for level flight, but it remains to remind you that "level" flight means keeping the wings level, not just aiming at a distant point and wallowing over there. A drooping wing makes the airplane want to turn, and to hold straight with a wing down means using rudder pressure to hold a course. The result is sloppy flying, like skidding on ice.

To overcome this bad habit, look out at both wingtips, while sitting on the ground, and see where they are in relation to the horizon (above it in high-wing aircraft, below it in low-wing ships). Then, in flight, use the same reference points to maintain lateral level.

There is also a tendency in making turns in a side-by-side

aircraft for the pilot, flying in the left seat, to gain altitude in one direction and lose altitude in the other. The reason is that in a right turn, the horizon appears lower than it does in a left turn. Again, the answer is a matter of selecting the proper reference points.

Similarly, in climbs and glides, the nose of the aircraft will appear above and below the horizon respectively, how much depending on the type of aircraft, how high the pilot sits, etc.

It may be well to mention here that there is no reason why a pilot learning to fly by VFR should not borrow from IFR techniques to operate his craft with more precision. The instruments are there; why not use them? This method of learning is called the *Integrated*

Fig. 5-4. Artificial horizon may be primary instrument in checking attitude of aircraft in flight. (courtesy FAA)

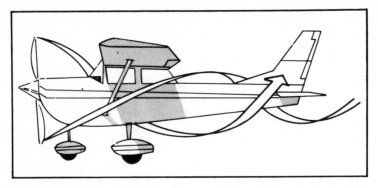

Fig. 5-5. Propeller slipstream (propwash) strikes rudder more strongly on one side than the other. (courtesy FAA)

System, and is becoming more and more popular with CFIs. By referring to the artificial horizon (Fig. 5-4), it is easy to tell when the aircraft's attitude is right or wrong. Later on we will discuss instrument flying techniques and show how many other flight instruments can be employed as cross-checks—airspeed, rate of climb, turn indicator, etc.

Your instructor will demonstrate for you another force that comes into play during climbs. It is called *torque*. Your vertical stabilizer is preset by the manufacturer to maintain a straight heading in cruise flight, even though the slipstream is actually spiraling back from the propeller so that it strikes the left side of the fin with more force than the right side (Figs. 5-5, 5-6).

In a climb, however, you are imparting more energy to the slipstream by flying with more power, and at the same time climbing at an airspeed lower than cruising speed. The combination of these two factors makes your aircraft want to swing to the left; hence a certain amount of right rudder pressure must be carried in climbs (or the effect trimmed out, if your aircraft has a rudder trim).

In descents, you will practice two different ways of losing altitude—gliding down the sky with power off, and descending at a recommended airspeed and power setting to give you a rate of descent of between 400 and 500 feet per minute. In power-off glides, you will learn experimentally the airspeed that gives you the flattest glide angle (this will vary with the aircraft's gross weight and with different gear and flap settings).

It is important to recognize that control forces are far different in power-off gliding turns than in normal turns under cruise power. Absence of the usual slipstream and slower airspeeds give an entirely different "feel" to the controls. It's good insurance to learn to make coordinated gliding turns, particularly near the ground on landing approaches.

Slow Flight and Stalls

We have already discussed at length what a stall is and how it can be anticipated with an angle of attack indicator, but until this new instrument comes into general use, you're going to have to demonstrate to the FAA that you can keep your airplane flying just above a stall with only such secondary instruments and reference points as your airspeed indicator, the position of the nose, and kinesthetic "feel" in slow flight.

Fig. 5-6. The vertical stabilizer on the Temco Buckaroo is deliberately offset to compensate for strong propwash torque.

Some modern airplanes with well-designed wings will virtually "hang on" at amazingly low speeds, but the stall is not the only danger you must look for. Equally important is the *sink rate*, or how fast you are losing altitude, flying nose high. It's just as bad to drop onto the ground without even stalling as it is to stall and then drop. Hence, by carefully watching your altimeter and rate of climb (or descent) indicator you can learn to maneuver at minimum airspeeds without losing altitude.

In slowing from cruising airspeed to approach airspeed, your instructor will show you how to smoothly reduce power *below* approach power setting, holding back pressure on the elevator control until the speed reduces to approach speed, then reapply approach power to start downhill under control.

Mastering slow flight will make you a better and safer pilot, and you will discover that by so doing you are widening your operating range of airspeeds in a way that will build confidence and prepare you for such advanced maneuvers as *stall down* landings.

Not for the novice, stall down landings require precise airspeed control very close to a stall. Properly executed, they will enable you to settle down over a high obstacle without flaps and without slipping, so that your landing actually can be effected with no flareout. At the last moment before touchdown, you "sink" your stick full back, gaining the last bit of extra lift and touching the runway as light as a feather on the cushion of air under your wing. In a slight headwind, expert pilots can land a ship this way with brakes locked!

Coordination Exercises

Beginners and professional pilots alike keep sharp in flying technique the same way pianists and violinists do—by constant practice. Just as finger exercises are a must for musicians, so coordination exercises are the simplest means of maintaining aircraft control proficiency.

Corporate aircraft pilots and air transport drivers sometimes find frustrating the monotony of just sitting there, flying straight and level, letting their autopilots do their work for them. On weekends, a growing number of professional airmen may be found doing aerobatics in planes like the Citabria, and lazy eights and chandelles in other lightplanes. These are advanced coordination maneuvers, highly effective in forcing the pilot to combine constantly changing airspeed, heading, and bank using the full three-torque system of aileron, rudder, and elevator (Figs. 5-7, 5-8).

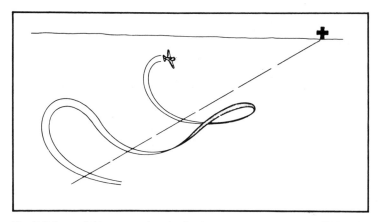

Fig. 5-7. Lazy eights are gentle coordination maneuvers. (courtesy FAA)

The simplest coordination exercise and one of the most effective in helping you learn smoothness in flying is the Dutch roll, in which you hold the nose dead on a point on the horizon while gently rocking the wings. It's not as easy as it sounds, for it demands coordination of rudder and ailerons just the reverse of coordination in normal turns.

Another good "finger-and-toe" exercise is simply to make shallow or medium turns left and right, of anywhere from 30 to 90

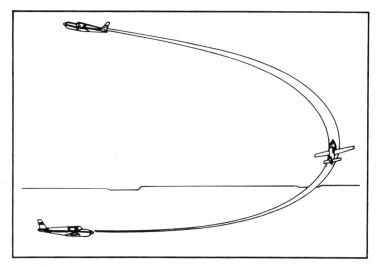

Fig. 5-8. Chandelles are basically maximum performance climbing turns with a 180° change in direction, the aircraft varying from maneuvering speed to just above a stall. (courtesy FAA)

degrees, while progressing toward a distant point. All such exercises can be done while flying from the airport to the practice area and back, rather than sitting there killing time watching the cars go by below.

If you have a radio with a broadcast-band receiver, there's no reason why you can't tune in a little rock and roll and literally "dance" across the sky in time to the beat! Just because it's fun doesn't mean its bad for you. In fact, rhythm in flying will make you a superior pilot to the guy who flies mechanically, so get out there and boogie!

Ground Reference Maneuvers and Traffic Patterns

We'll be returning to this topic later because there's a lot to say about the proper way to fly an airplane in relation to fixed ground objects in still air and on windy days. Suffice it here to say that the general idea of primary ground-pattern maneuvers is to teach you how to control your aircraft in airport pattern flying, and later on in other low-flying situations.

Low flying, incidentally, has earned a bad name because there are some pilots who can't resist showing off, buzzing friends' houses and otherwise violating not only the Federal Aviation Regulations but the laws of decency, common sense and survivability.

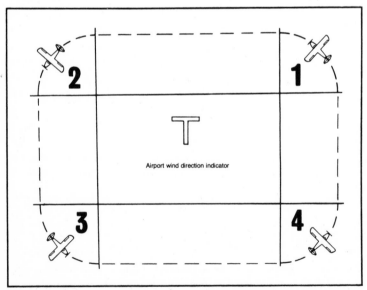

Fig. 5-9. Flying a rectangular pattern around a field teaches you to fly with reference to fixed landmarks while allowing for wind drift.

All that the FAA requires in low flying over sparsely settled non-congested areas or open water is that you always have a place to land in an emergency without endangering persons or property on the ground, and that you stay at least 500 feet away from people, boats, vehicles, and structures. Crop dusters, oil line survey pilots, and other professionals whose airwork demands that they fly low usually do so with waivers.

While emergency landings are not now required as part of your private pilot flight test, the instructor who does not teach you how to make them may not have your best interests at heart. Incidentally, the main reason why the FAA dropped forced landings from private flight tests is reasonable: You can learn how to judge gliding distance at your home airport, and too many accidents have happened while simulating emergency landings.

In simulating traffic pattern flying, using a square field bounded by hedgerows or roads (Fig. 5-9), you begin to learn how to make the airplane go where you want in relation to fixed landmarks. Or, as some pilots put it, how to fly "with your head outside the cockpit."

Takeoff and Departure Procedures

In Chapter 4 we discussed some of the basic problems of takeoffs, and while this maneuver is relatively simple, it may still present the greatest hazard in a flight. Just as in approaching to land, the pilot is preoccupied with two kinds of motion, transitioning between two-dimensional travel on the ground and three-dimensional flight.

On the takeoff run you steer with the rudders, but the moment you're airborne, the rudders are used to offset torque and offset aileron drag as explained earlier (Figs. 5-10, 5-11). You have scanned your instrument panel prior to beginning your takeoff roll, then shifted your vision to infinity to hold straight. You may sneak a look at your tachometer to see if the engine is turning up maximum power, but this is distracting. Instead, your ears are tuned to the throbbing of the exhaust, your fingertips sensing the building pressures as the air became "hard."

Different aircraft behave differently on takeoff; each has its best pitch attitude for liftoff, one that permits the most rapid development of airspeed and control effectiveness. This is what the pilot "feels" for, a resistance to control movement that is not a measure of speed but rather of controllability.

Trouble sometimes develops when an overanxious pilot rotates into climb attitude too soon, forcing his aircraft off the runway

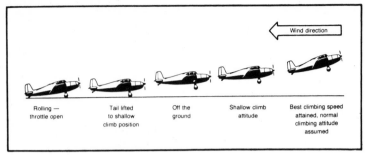

Fig. 5-10. Standard takeoff. (courtesy FAA)

below V_{mc} (minimum control speed). When that happens, he may find himself settling back onto the runway with more drag than lift tugging at him, holding him earthbound.

By taking off at an attitude that produces the best rate-of-climb airspeed and continuing under full power to about 400 feet altitude, the pilot puts himself in the best position to maneuver safely in case of engine failure. Use of full throttle also rives a richer mixture in many engines, providing additional cooling for climbout.

One of the big surprises most pilots experience on their first solo is the way their lightened trainer seems to leap into the air after their instructor climbs out. There is nothing abnormal about this; in fact, it is an enjoyable experience.

Approach and Landing Procedures

Getting an airplane back onto the ground is frequently regarded

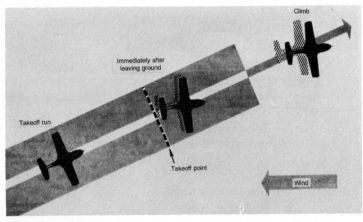

Fig. 5-11. To compensate for a crosswind on takeoff, crab into the wind when airborne sufficiently to offset drive. (courtesy USAF)

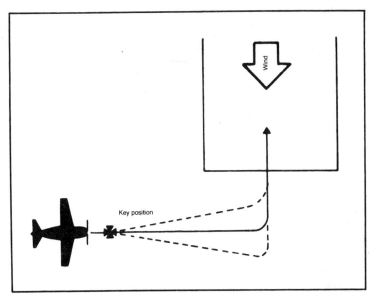

Fig. 5-12. In a 90° approach to landing, set up key position on base leg approximately 45° from intended landing point. You can then vary base leg to compensate for wind conditions. (courtesy FAA)

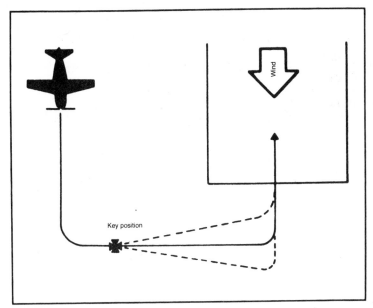

Fig. 5-13. On 180° side approach to land, pilot makes two 90° descending turns, onto base and final, using 90° key position to handle wind correction. (courtesy FAA)

by fledgling pilots as the most difficult part of flying; once you have mastered that, you've learned all there is to know!

As a matter of fact, the landing is the last in a sequence of events that begins at what is called the *key position* in the traffic pattern (Figs. 5-12, 5-13). While there are other kinds of approaches to landings—the straight-in and the 360-degree overhead approach are two—in normal practice you will begin your descent from pattern altitude either on the downwind leg opposite the landing point or on the base leg at a point about 45 degrees from touchdown.

You will have turned on carburetor heat to prevent icing in a power-off glide, reduced your airspeed with elevator back pressure to your desired approach speed, and then either trimmed the ship for a power-off gliding approach or reset the throttle to control the rate of descent. This latter method is preferred, for you can more easily maintain a constant glide angle through any updrafts or downdrafts with small power changes.

Some pilots prefer to come in high, then cross the field boundary at a steepened angle by using flaps or closing the throttle (Fig. 5-14). This is considered good "insurance" against running out of altitude in the event you encounter a sudden subsidence or unexpected lower headwinds.

To determine in advance whether you are gliding safely over the telephone wires you find around most all airports or other

Fig. 5-14. This Bonanza pilot has gear and flaps down for steep approach to land.

Fig. 5-15. In approach over telephone poles and wires, watch to see that top of pole lowers below horizon.

obstacles such as trees or rooftops, use a simple trick of perspective. Watch what the top of the obstacle does in relation to the ground behind it (Figs. 5-15 through 5-17). If the telephone poles "grow" taller, you're undershooting. If they drop down in your field of vision, you've got it made. If they stay put, chances are you will be mistaken for a bird trying to land on the wires!

Fig. 5-16. This Bonanza pilot makes steep flap approach over trees into short field in Sierra Nevadas.

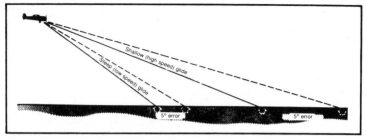

Fig. 5-17. A shallow glide provides more room for error than shallow one. (courtesy FAA)

Once you have set up a good glide angle, make sure nobody is sneaking in under you on a long flat approach, line up with the active runway, and watch for side draft. A crosswind from your right will drift your plane to the left, and vice versa. To correct: either stick a wing down on the upwind side, holding the nose straight with the rudder and letting a sideslip compensate for drift, or, better still, crab into the wind sufficiently to maintain a straight track into the middle of the runway. In the latter case, the crab may be held to the moment just before touchdown, at which point you kick the nose straight and in so doing provide a skid input that offsets wind drift and holds you on track at the critical second of landing (Fig. 5-18).

The landing actually comes after one more element in the approach, called the *flareout* (Fig. 5-19), in which you reduce your approach airspeed to stalling speed, getting every last ounce of lift out of the wings so that when the aircraft touches down it is finished flying and stays down (this is true in both taildraggers and tricycle gear landings).

Back on the ground, you're not through yet. You still have to hold straight during the landing roll until you have slowed sufficiently to turn off the runway. In a crosswind, your ship will want to

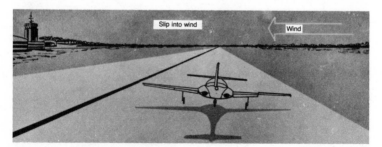

Fig. 5-18. Crosswind touchdown slipping into wind to maintain centerline position on runway. (courtesy USAF)

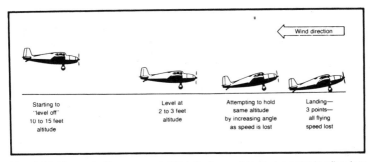

Wind direction

Starting to
"level off"
10 to 15 feet
altitude

Level at
2 to 3 feet
altitude

Attempting to hold
same altitude
by increasing angle
as speed is lost

Landing—
3 points—
all flying
speed lost

Fig. 5-19. Flareout after approach kills off airspeed for three-point landing into wind. (courtesy FAA)

weathercock (if it's a taildragger) so stay alive on the rudders, your vision again fixed on infinity to help you roll straight.

Crosswind landings are much easier, in aircraft with tricycle gear, because the weathercocking tendency is offset by the tendency of the aircraft's center of gravity, ahead of the main gear, to remain in motion straight ahead. This is according to a law of motion dreamed up by Sir Isaac Newton, who, although not a pilot, knew pretty well what he was talking about.

There are many fine points, not only in making good landings but in the full flight spectrum, that your instructor will tell you about as you progress toward solo, so we'll let him earn his money while we go on to more advanced airwork in the next chapter.

Chapter 6
Your First Solo Flight

Talk to any 15,000-hour airline captain today and ask him what was the greatest thrill he has had in his flying career. Aside from having his plane hijacked and flown to Cuba at gunpoint, riding out a "blue storm" of clear air turbulence (CAT), or sharing his cockpit with a sexy movie star on a charter flight, he'll probably say: "My first solo!"

The moment your flight instructor steps out of the cockpit, grins, and says "It's all yours!", you are a different person. At first the cockpit appears terribly empty; a feeling of loneliness overwhelms you, and you resist an urge to cry out "Hey! Come back! Don't leave me alone!"

Until this moment you have been tied to your CFI by an umbilical. He is always there to keep you out of trouble, no matter what you do wrong. But suddenly your life depends on your own skill, and you have yet to put it to test. Can you *really* fly—all by yourself? Or is your instructor just trying to get rid of you by letting you go up alone and kill yourself?

You swallow hard and taxi away from him—sitting there nonchalantly on a fence railing, lighting a cigarette and watching you move off, a spectator at a Roman circus watching a Christian being tossed to the lions. Well, you'll show him! You'll fool him by making a perfectly sensational circuit and landing. Lindbergh . . . Von Richthofen . . . Orville Wright . . . what did they have that you don't? And so off you go, into the wild blue yonder . . .

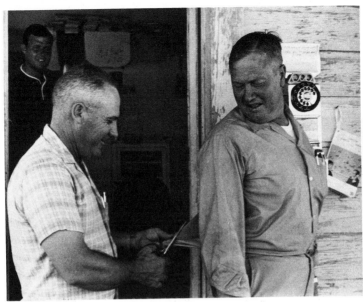
Fig. 6-1. Your first solo calls for ceremonially snipping off your shirttail to nail on bulletin board.

Curiously enough, there is a growing belief within the ranks of FAA General Aviation Field Offices that the solo flight is an unreal goal, a false plateau that ought to be eliminated. It just doesn't make sense, these non-believers contend, that a pilot barely able to get around the traffic pattern and back onto the ground in one piece should be turned loose with a lethal machine. Rather, he should fly dual all the way to the point where he qualifies for his private pilot license!

"You don't turn loose a student motorist until he is ready for his driver's license," they point out. "Cheez, can you imagine a bunch of solo student drivers careening along the freeway *practicing?*"

There is a lot of logic in this argument, but until the FAA changes the rules, the first solo will remain a memorable milestone in the life of any pilot—a sort of puberty ritual one goes through to prove himself a man (Fig. 6-1). In fairness to tradition, it has worked pretty well in the past, and while student pilots today are exposed to far more procedural flying than they were 20 years ago, there is virtue in giving a fledgling birdman confidence to go ahead and complete his flight training. Nothing builds confidence better than a good first solo flight.

There are certain new requirements to prepare you for that

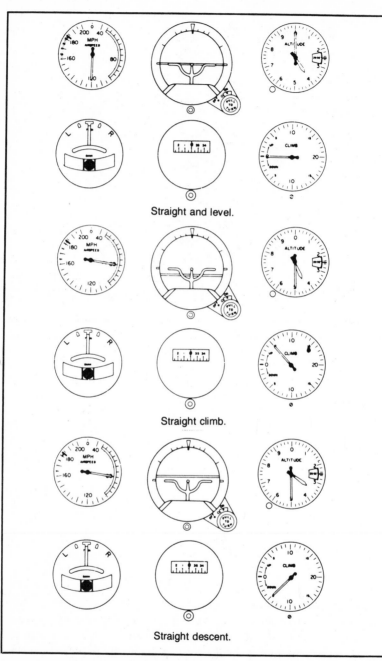

Straight and level.

Straight climb.

Straight descent.

Fig. 6-2. Flight instruments are scanned, using the artificial horizon as primary instrument, others as supportive. (courtesy FAA)

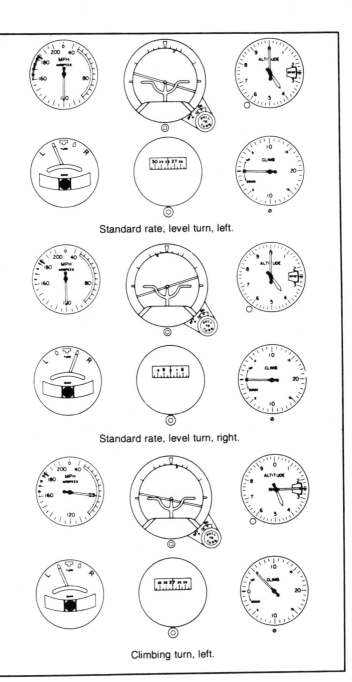

Standard rate, level turn, left.

Standard rate, level turn, right.

Climbing turn, left.

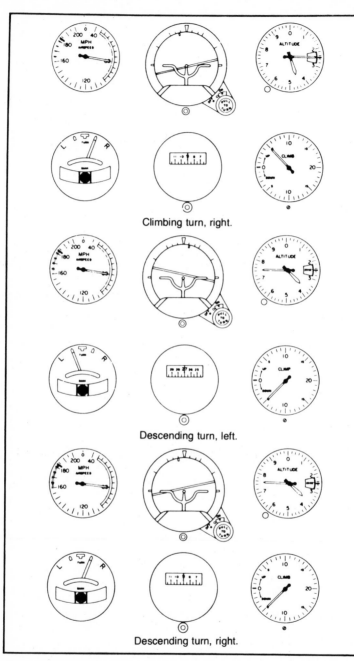

Climbing turn, right.

Descending turn, left.

Descending turn, right.

Fig. 6-3. Instruments are grouped around primary instrument, the artificial horizon. (courtesy FAA)

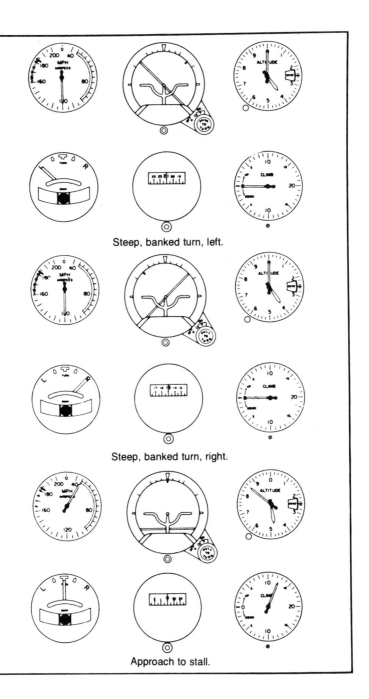

Steep, banked turn, left.

Steep, banked turn, right.

Approach to stall.

solo, although, oddly enough, minimum dual instruction is not one of them. In the FAA's Advisory Circular AC 61-2A, the *Private Pilot (Airplane) Flight Training Guide* (an excellent manual designed to assist independent CFIs in planning student flight lessons), the first solo is tossed in arbitrarily as a part of Lesson 8. The presumption is that the CFI gives his students eight one-hour lessons, and it says: "At the completion of the dual portion of this lesson, the student should have achieved reasonable proficiency in all the flight training maneuvers he has received, be able to make safe takeoffs and landings consistently without assistance or direction, and recover from poor approaches and bad bounces. He should have demonstrated the ability to solve all ordinary problems to be encountered on local flights."

The entire flight training syllabus in AC 61-2A, in fact, is based on the *integrated* system of flight instruction introduced to the Civil Aviation Regulations by amendment back in 1960. This is one of the most logical changes in flight instruction techniques since Orville said to Wilbur: "If you let go of that wingtip, maybe I can get this thing off the ground!"

By *integrated* instruction, the FAA is talking about the student performing each maneuver by both IFR and VFR techniques, using outside visual references and flight instruments alternately from the first time the maneuver is introduced. It's up to your instructor whether you first do a maneuver, say a medium-banked turn, by watching the horizon or the instrument panel. The point is that by doing them both ways, you integrate two sets of references and learn to use both, together or interchangeably. This builds confidence in flying by instruments alone, because in your mind's eye you get a better picture of what is really happening (Figs. 6-2, 6-3).

In practice, the student wears a flip-up visor which when lowered acts like a set of horse blinders to prevent him from seeing outside the cockpit when flying on the gauges alone. This permits him to concentrate on one set of references with the hood down, another with the hood up.

The idea is not new. During World War II, I instructed British cadets at Falcon Field, Arizona, under the Empire Training System, which pioneered the integrated system by requiring IFR practice on every pre-solo flight. Unquestionably it helps the student pilot achieve greater precision and competence more quickly, particularly in holding altitude in level flight and in turns, controlling airspeed in climbouts and approaches to land, and holding a compass heading on cross-country flights.

At the beginning of Chapter 5 we listed 26 maneuvers and procedures required by the FAA prior to going up for your check ride for your private pilot license, and we discussed ten of them, the basic things necessary to completion of a flight around the airport traffic pattern from takeoff to landing. Now let's continue on down the list of requirements of things you'll have to learn prior to your first solo flight.

Stalls from Critical Flight Situations

If there is any flight maneuver more misunderstood and yet more vital to safety than a simple wing stall, I have yet to hear about it. The confusion stems from the difficulty in mentally grasping just what *critical angle of attack* is—a stall results solely from attempting to fly above it, at an excessive angle of attack.

To begin with, let's make it clear what angle of attack is *not*. There is a popular misconception that the angle of attack is the angle between the wing's chordline and the Earth's surface; this is true *only* if the direction of flight is parallel to the Earth's surface.

The angle of attack is solely a measure of the angle between the wing chordline and the direction of flight, and it is an invisible thing. An aircraft sinking with its wing chord level may still exceed its critical angle of attack and stall, while another aircraft zooming straight up may be safely flying at zero angle of attack (Figs. 6-4, 6-5).

The introduction today of a variety of angle of attack indicators for general aviation helps to make the angle of attack "visible," yet the FAA is cautious about endorsing their use wholeheartedly. "The Navy has used them for 30 years," one FAA flight test engineer told me, "and if there were foolproof indicators available, don't you think we'd adopt them?"

The problem is that many things enter into the formula defining *critical angle of attack*, the most important being airspeed, gross weight of the airplane, and the load factors imposed by maneuvering. Center of gravity, flap configuration, dirt, and moisture can also affect an angle of attack indicator's reading. I have flown a Beech C33 Debonair equipped with an angle of attack indicator of the type used by Ohio State University's Aviation Department in an FAA-sponsored flight training program, and found it revealing.

The thing to remember is that you can stall an airplane *at any airspeed* simply by exceeding the critical angle of attack. Applying

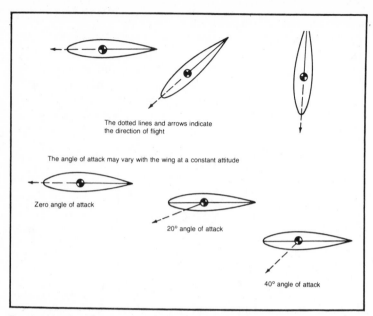

Fig. 6-4. Angles of attack of a wing in flight. (courtesy FAA)

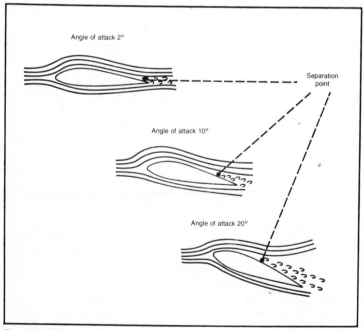

Fig. 6-5. Effect of angle of attack on airflow separation point. (courtesy FAA)

abrupt or excessive back pressure on the elevator control at a relatively high airspeed can result in what is called an *accelerated* or *high-speed stall*.

The Air Force's Air Training Command states this more succinctly in its ATC Manual 51-4 (*Primary Flying, Jet*): "Basically there is one cause for a stall—high angle of attack . . . It is important to realize that an airplane can stall at any airspeed, attitude, or power setting if you demand with the elevators an angle of attack above the critical value" (Fig. 6-6).

Once you get it straight in your head what a stall is and understand that it is the critical angle of attack and not the airspeed alone that causes it, then it's time to discover by practice and demonstration why they are most dangerous when encountered in critical flight situations—during takeoff and departures, approaches and landings, and while executing abrupt maneuvers anywhere in the sky.

An increase in the gross weight of your aircraft, or in load factors imposed during steep turns and other flight maneuvers will increase the wing loading and so increase the indicated airspeed at which the critical angle of attack is reached.

Thus, prior to solo, you will practice getting into trouble (and out of it) by simulating stalling conditions at an altitude safe for recovery. This is so if you goof up at low altitude on takeoff or landing, or in zapping around to miss somebody, you will know that you have simply reached your wing's critical angle of attack, and you know what to do about that!

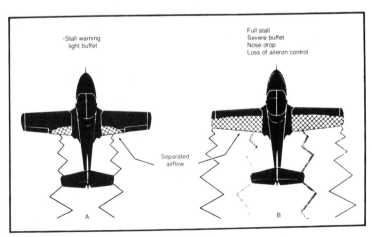

Fig. 6-6. Stall begins at wing root with light buffet, extends to wingtips in full stall. (courtesy USAF)

If you've forgotten, the FAA wants you to do it their way, which is:

1. Reduce angle of attack with the elevator control.
2. Attain straight and laterally level flight by coordinated use of flight controls.
3. Apply smoothly all available power.

Once your CFI has demonstrated and let you practice simulated stalls in critical flight situations, it will become almost instinctive for you to dive at the ground instead of shying away from it when you think you're about to fall down. It's a new instinct that may someday save your life.

Steep Turns

Before your instructor will sign off your student pilot ticket with an endorsement that you are ready to solo, he will make sure that you know what a steep turn is and what its inherent dangers are. You really won't need to make steep turns, unless an emergency demands it, but should this occur you should know what to expect. Later on, you'll learn to do them expertly as a basis for advanced maneuvers.

Remember first that load factors, in terms of gravity forces, double in a 60-degree banked turn. At 80 degrees of bank you will exceed a load factor limit of 6 Gs, the limit of a good aerobatic plane. A level 90-degree banked turn is mathematically impossible (Fig. 6-7).

Don't forget too that an airplane's wing will reach its stalling angle of attack at a higher airspeed in a steep turn than in level flight. Mathematically, the stalling speed increases as the square root of the load factor. This means that in a 60-degree bank an aircraft that normally stalls at 60 mph in level flight will reach its critical angle of attack at nearly 85 mph.

Because of the nice coordination required to do a good steep turn, it becomes a fine introduction to advanced maneuvers that require swift, sure analysis and a fluid coordination (Fig. 6-8). In learning under the integrated method of primary flight instruction, reference to the ball-bank (often called the slid-skip indicator today) will quickly tell you what you will soon learn by kinesthetic sense ("feel")—how well you are coordinating stick and rudder pressures.

For the novice, it is well to understand here that one of the deadliest maneuvers you can get yourself into flying on instruments is the spiral dive, which pilots call the *graveyard spiral* for very good reason. Improper correction for overbanking tendency and poor coordination in a steep turn permits the nose to drop and airspeed to build up with amazing rapidity.

With no outside horizon for reference, your senses will deceive you at this point, the sensation within the inner ear convincing you that you are turning in a direction opposite from what is actually going on. Before artifical horizons came into general use, a confused pilot in a spiral dive might attempt to correct the sudden runaway airspeed by hauling back on the stick, as he would do in level flight—after all, he was trained (properly) to regard his elevators as the airspeed control! In a spiral dive, however, back

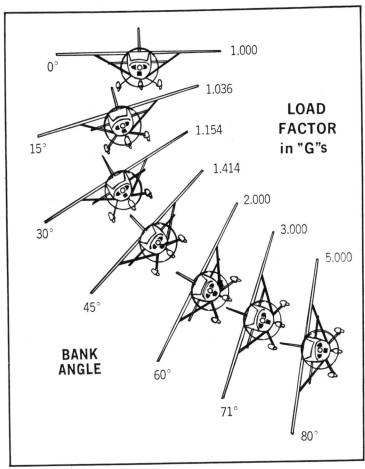

Fig. 6-7. Load factors double in 60° banked turn, multiply five times in 80° bank. (courtesy FAA)

Fig. 6-8. Good steep turn requires nice coordination.

pressure only tightens the turn until the aircraft's ultimate load factor is reached and the wings pull off, or he lands in somebody's petunia patch.

Fortunately, artificial horizons show what is really happening in such a situation at a glance. By cross-checking with other instruments, the pilot gets the true picture, irrespective of what his senses are telling him. A high airspeed, turn needle widely deflected, compass spinning, and wild rate of descent all shout what your artifical horizon is trying to tell you—you're in a spiral dive, going down like a bat into Hell. Recovery is simple, fortunately. Take off power, roll your wings level, and bring up the nose to level flight.

While on the subject of steep turns, let's dispel another old pilots' tale that says you can always tell when you have completed a perfect 360-degree circle: You run into your own "propwash" and feel a definite bump. What you feel actually is *not* propwash but vortex turbulence, which trails downward behind your wingtips. If you run into your own vortex, you've lost some altitude.

Later on, after solo, when you get the time to work at polishing up those steep turns, you will discover that more rudder pressure is necessary to enter a right turn than a left turn, due to something called *torque effect*. This happens due to three things: You've applied extra power to maintain adequate lift in the turn; your airspeed drops below cruising; and your tighter (shorter radius) turn makes gyroscopic forces in the propeller-crankshaft assembly stronger.

You will also notice that it takes less corrective aileron pressure to counteract overbanking in right turns, and that there is a tendency to slip in right turns and skid in left turns as you seek to establish proper rudder pressures to fight torque. Incidentally sneaky fighter pilots in World War I found out that if they attacked from 9 o'clock high and forced enemy pilots into a steep left climbing turn, the gyroscope effect of their foes' rotary engines caused them to stall out. The same phenomenon was a decided danger in steep turns after takeoff, for obvious reasons.

Crosswind Takeoffs and Landings and Slips

This area of discussion will raise a lot of blood pressure among CFIs, for no two instructors seem to agree on the best way to handle an airplane in a crosswind. Takeoffs they generally agree on, but landings? Ugh!

If the wind always blew right down the midde of the runway, there'd be no problem of transitioning from ground to sky operations and vice versa. But more often than not, a crosswind component is present just when you don't need it. Actually, there is no mystery about handling an aircraft in a crosswind, as long as you do not attempt to take off or land with a crosswind factor greater than the maximum allowable for your aircraft.

On takeoff, whether you are flying a taildragger or a tricycle gear aircraft, what you are attempting to do is build up flying speed while still ground-bound, driving down the centerline of the runway while the wind is trying to force you off to one side. To resist the crosswind effect, simply hold your ship on the ground a little longer than you would in a normal takeoff, letting your ground speed build to the point where your wing is already flying. Then lift off the runway quickly to prevent a sideways skittering that can impose side loads on your landing gear.

During the takeoff run, many pilots like to hold the stick into the wind to reduce lift of the upwind wing (which wants to fly prematurely), removing the aileron pressure as they become airborne. At this point, free of the ground, a crab angle is set up by turning into the wind just enough to maintain a straight climbout.

The same set of forces are at work in a crosswind landing. You are trying to set down on the runway centerline with your aircraft pointed the way it will roll once on the ground, while the crosswind tries to drift you off to one side.

For some reason I cannot fathom, many pilots prefer to fight the crosswind on a final approach by aiming the airplane straight and

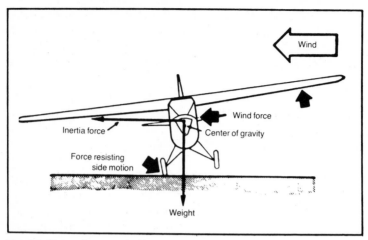

Fig. 6-9. Forces in crosswind landing. (courtesy FAA)

sideslipping into the wind to counteract drift. The result is an abnormal, uncomfortable, and (to me) totally unnecessary bunch of hard work to achieve something you don't need until you're ready to land.

"But my goodness!" cry advocates of the "tilt" school. "Why wait until the last moment to try to straighten the airplane and land it at the same time?"

I reply: "Why not?"

In actual practice there is less work involved in simply crabbing down the final approach, maintaining a straight track over an extension of the runway centerline through unpredictable wind shears than sitting sideways, cross-controlling and spilling martinis back in the passenger cabin. While you're still airborne, you don't have to "aim" the airplane to make a clean approach; simply fly it in a normal fashion—sideways in a stiff crosswind if necessary!

The same holds true, if you believe me, during your flareout. Not until you are ready to put the ship on the ground is there any logical reason to point it straight ahead. As a matter of fact, there is a very good reason to wait until you're ready to set down on the runway to do this, because, as you are flying with wings level and nose crabbed into the wind, all it takes is one swift kick on the downwind rudder to straighten out and land right. This same rudder movement at the same time skids you into the wind just enough to compensate for drift, and there you are (Fig. 6-9).

Once again, when you're on the ground, hold it there to prevent skittering; with a taildragger, don't forget the weathercocking force that wants to turn you into the direction from which the wind is coming. Say the wind is from the right; you've crabbed down the approach, holding dead on the runway centerline, flared out still crabbing, and then booted left rudder to straighten out, skid into the wind and land all lined up pretty. Your left foot is already in action, toes curled around the rudder pedal, holding against the weathercock tendency.

There's really no difference whether you prefer to crab down the sky and straighten out at the last moment, or use a sideslip into the wind all the way down the sky, if you want to work that hard. But the FAA seems to want you to at least be able to demonstrate the slip approach, so let's take a look at it.

There are actually three kinds of slips—sideslip, forward slip, and slipping turn—but all are executed about the same way, except

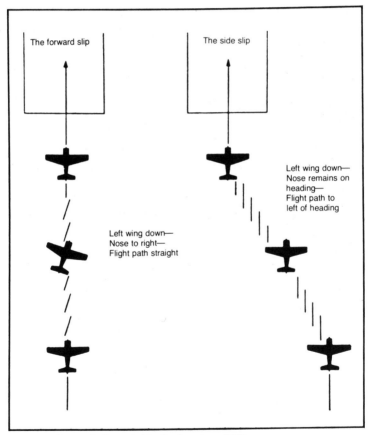

Fig. 6-10. Forward slip and sideslip. (courtesy FAA)

Fig. 6-11. Aeronca 7AC pilot executes steep forward slip on high approach to land.

in degrees (Fig. 6-10, 6-11). A sideslip is done by putting one wing down in a bank and holding the nose straight with opposite rudder. In a crosswind, this tracks you straight ahead, if your slip is equal to drift. In a forward slip you simply crab to one side or the other, then crosscontrol. You're doing exactly the same thing as you did in a sideslip, only your ground track is different. In a slipping turn, all you do is hold top rudder and extra back pressure and let her go around.

There are two reasons for slipping. One, as we discussed, is to track over the ground the way you're pointed in a crosswind. The other is to increase your rate of descent. On a final approach into a shortfield, in lieu of flaps you can use any or all three types of slips, smoothly cross-controlling and not worrying about stalling—you can't hold an airplane in a slip at stalling speed.

Short- and Soft-Field Takeoffs and Landings; Maximum Climbs

If everybody flew from great big airports with no obstructions at the ends of the runways, there would be little reason to concern yourself with short-field takeoffs and maximum climbouts. But much of the value of a light aircraft lies in its ability to operate from small airports, and there is an old pilot's adage that says you can always find a small airport by following the power lines to it.

You don't have to fly like a V/STOL (vertical/short field takeoff or landing) pilot, but you should know how to get in and out of little fields safely. Who knows—maybe you'll be flying up to a mountain meadow airport beside a crystal trout stream, and it would be a shame not to know how to land and take off when the fish are biting!

A short-field takeoff procedure assumes a firm, smooth surface for the takeoff run, in which you want to achieve minimum drag to get airborne fastest. Apply full power smoothly and quickly and, if you are flying a taildragger, get your tail up as soon as the elevators become effective (to reduce wing drag). The flap setting used should be that recommended by the manufacturer in the Airplane Flight Manual for shortfield takeoffs.

Some pilots claim it's a good idea to set the brakes and run up the engine to maximum obtainable rpm before starting the takeoff run. Others say definitely not; you gain nothing because your propeller cavitates (forms a vacuum around the blades) and also sucks up rocks that can pit its leading edges.

For steepest climbout, to get over those treetops at the end of the runway, hold the aircraft on the ground, rolling on the main wheels until you attain best angle-of-climb airspeed (V_x). Then lift off smoothly and climb straight ahead, in theory until you reach 50 feet altitude or enough to miss those bird nests. (Again, follow recommendations in the Airplane Flight Manual for best airspeed, power setting, flap configuration, etc.)

Just in case you're flying a multi-engine craft, climb out at minimum engine-out control speed (V_1) so you won't lose control if one engine quits unexpectedly.

For all normal climbs, a slightly higher airspeed called the best rate-of-climb (V_y) is used, unless the Airplane Flight Manual recommends a still-higher climb speed to give you better engine cooling or pilot visibility. In short-field takeoffs, then, you'd use best angle-of-climb (V_x) up to 50 feet or so, then continue at best rate-of-climb (V_y) to cruise altitude.

In soft-field takeoffs, your concern is not with wing drag but *wheel* drag from mud, slush, snow, or soft dirt. Obviously, the thing to do is shift the aircraft's weight from wheels to wings as quickly as you can, simply by maintaining a nose-high attitude through the takeoff run.

By holding this nose-high attitude, you'll lift off at an airspeed slower than V_x because of a thing called *ground effect*, best described as a temporary gain in lift due to compression of the air "cushion" beneath your wings when flying near the ground (Figs. 6-12, 6-13).

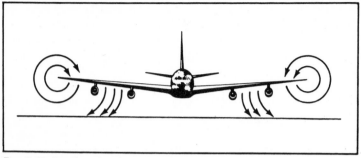

Fig. 6-12. Interference of ground with tip vortices and resulting downwash. (courtesy USAF)

Once airborne and out of the goop, smoothly reduce the angle of attack to either V_x or V_y climb speed for steepest or fastest climb-out.

In small training aircraft, incidentally, you may not notice much difference in climb performance at varying airspeeds, but in higher-performance craft that cruise at 150 mph or better—and particularly in multi-engine craft—it becomes critical to get the airspeed just right.

Short-field landings are simple to execute, if you follow procedures in the Airplane Flight Manual regarding power setting, airspeed, and flaps. Normally you'll be using full flaps, gear down,

Fig. 6-13. Landing Cessna 170 flies in ground effect in flareout at Saline Valley, California.

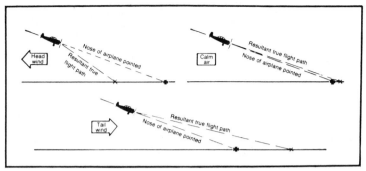

Fig. 6-14. Effect of wind on landing aircraft. (courtesy FAA)

and an airspeed 1.3 to 1.4 times power-off stall speed (between 78 and 84 mph for an airplane that stalls at 60 mph).

Once your airspeed is set, the use of power to control your descent with a constant pitch attitude is quite simple, if you heed one warning: In a steeper angle of descent, you may find a tendency to break your glide and flare out to land *too high*. There is an optical reason for this, having to do with a "break" in perspective that normally tells a pilot it's time to level off. The steeper the approach angle, the higher this visual change in perspective occurs (Fig. 6-14).

Regardless, touch down at minimum controllable airspeed (V_{mc}) with the attitude about that of a normal poweroff stall. Nosewheel-type aircraft should be held in this attitude as long as possible during the landing roll, and the taildragger held in three-point attitude during braking for greatest drag.

Soft-field landings call for about the same procedure as short-field landings, with a word of caution: Try to land with the lowest possible airspeed and keep the nosewheel of a tricycle-gear craft up as long as possible. In low-wing ships, use full flaps with caution; they may be damaged by mud, rocks, or other debris kicked up by the wheels.

Power Approaches to Full Stall Landings; Wheel Landings

Steep descents with power are used for getting into a short field, but when encountering turbulence due to low-level wind shears or other factors, a power approach with a flatter slope is advisable because it permits quick corrections for up and down drafts. In such turbulent air, you will want to use a higher airspeed. Some pilots say "add 10 mph for safety and 5 more for good luck."

There is no reason to let your power-approach speed get too

high, only enough to stay above minimum control speed (V_{mc}) sufficiently to be able to handle unexpected gust loads with full control. Also, continue use of power through a wheel landing to provide control effectiveness until you are solidly on the ground.

In the steep, low-speed power approach into a short field, as we said, your control of attitude and airspeed will permit you to land with minimum flareout and no floating—a nice, precise way to set down in smooth air.

Power approaches in severe turbulence should be more shallow, terminating in a landing with the aircraft in approximately level flight attitude. Once the wheels touch, immediately close the throttle, holding the aircraft on the runway with forward pressure until the wing has stopped flying and you can safely get the tailwheel down (in a taildragger).

FAA examiners now encourage use of power during landing approaches under all circumstances. If properly planned, the throttle is slowly closed as the airspeed is reduced to stall speed at touchdown. This is good insurance against carburetor icing or engine stoppage caused by abrupt throttling.

Power landings, incidentally, take a lot of guesswork out of night landings on unlighted fields (or landing on glassy water in seaplanes), providing positive airspeed and rate-of-descent control all the way down.

Different flight instructors tailor their flight curricula not only to requirements of different students, but to varying environmental situations as well. If you are learning to fly from a large, smooth airport, early emphasis may be placed on radio communication procedures to help you fit into a busy traffic pattern. If your home base is a small, windy airport surrounded by power lines and tall trees (as too many are!), you must learn early to fly safely with appropriate departure and approach techniques.

In areas such as the Los Angeles basin, where smog frequently cuts visibilities to dangerous minimums, a knowledge of rudimentary instrument flying techniques is almost a must. And while you will normally have a good understanding of basic instrument attitude flying under an integrated flight training course, we will discuss instrument turns to headings and recovery from unusual flight attitudes and later on, where we can devote more time and space to it.

Similarly, your CFI will have briefed you early in your flight training on various emergency procedures—how to replace fuses or reset circuit breakers, crank the gear down in an emergency, or call for assistance by radio if lost or trapped above the clouds.

You will have learned how to abort a bad landing approach and execute a go-around, repositioning yourself into the normal traffic flow; what to do when the surface wind reverses; and how to handle your aircraft in the event of power loss near the ground.

However, it may be well at this point to discuss the most basic emergency procedure you should become familiar with prior to solo flight—the emergency or forced landing after complete power failure. Here is where all your newly learned skill comes into play, and where you show that you can *think* under pressure and plan a way out of your trouble.

To begin with, your instructor will have warned you never to get into a spot where you can't set down in case of engine failure. A pilot without an escape routine is living on borrowed time. In pre-solo forced landing emergencies, you will only be expected to be able to make a safe approach to land straight ahead or by turning up to 90 degrees.

The one basic rule in a forced landing is to get down on the ground in the *safest manner possible*. As said earlier, it's far better to fly into a tree dead ahead and shed a wing than to stall out and spin into the ground.

Always know where the wind is blowing from, either by checking an airport wind sock or traffic T or from natural indicators—cloud shadows, blowing smoke, ripples on water, cattle grazing with heads upwind.

Secondly, learn to estimate your glide angle, under prevailing wind conditions; once you have selected a field to land in, make sure you come in a little too high rather than a little too low. You can't stretch a glide; you *can* slip in and land short.

Color of the field tells you a lot about what to expect. Dark green means lush vegetation and probably soft, wet earth. Brown, hard earth is preferable. It may be better to land crosswind in the direction of windrows or plowed furrows. And there may be invisible power lines to trap you, so look carefully for power poles.

Once you know your wind, pick your field and turn toward it at a safe gliding speed. Settle down and fly smoothly, making a normal approach to land as you would at your home field. Once over the edge of the field you have chosen, use flaps or a forward slip and try for a good stall landing. Don't make the mistake of crowding your field and overshooting, or trying to dive at it. Don't blow your cool!

Chapter 7
Low Altitude Maneuvers

Most of your early flying lessons are devoted to getting familiar with your airplane, learning coordination and what effect control pressures have on attitude and performance. For safety's sake these are practiced at a safe altitude above the ground, where there's plenty of room to recover from stalls and spins and where you can devote full attention to the ship itself.

You first learn to refer primarily to the Earth's horizon for a reference line, one that appears to rise with you as you climb into the sky. When you become familiar with the feel of flying your aircraft, you begin to look down at the broad expanse of the world below you. Now comes the fun of picking out patterns of fields and roads that will be extra reference points when you begin flying through patterns of your own in the sky.

Wind Effect on Maneuvers

You are riding the wind as a boat rides the current of a moving stream, and wind drift complicates your maneuvering when you try to relate it to ground objects. Once you become familiar with this art of drift correction, a whole new world of flying fun opens up for you.

The first thing your instructor will show you is how the wind effect creates an optical illusion that must be taken into account in low-altitude pattern flying.

Let's go back for a moment to September 20, 1905, the day Wilbur Wright became the first pilot to fly through a complete circle

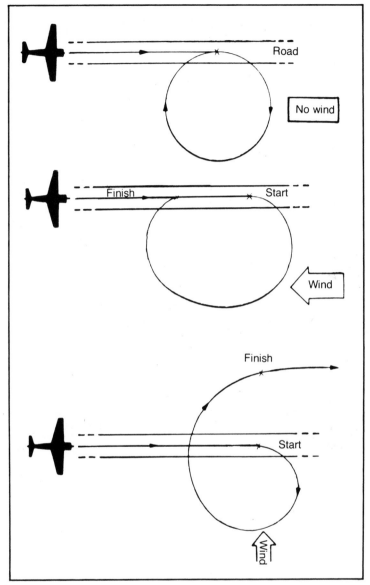

Fig. 7-1. Effect of wind drift in turns. (courtesy FAA)

over the ground, at Simm's Station, a trolley-stop near Dayton, Ohio. In a 90-acre meadow called Huffman Prairie, Wilbur took off in his little biplane, headed north, and then eased around in a left turn. Flying at treetop level, he suddenly picked up a fresh nor'-

easter that startled him. Drifting sideways, it seemed to Wilbur that he was slipping into the ground! Quickly he leveled his wings and landed, sure something was wrong with the airplane.

Later on the same day he tried it again, and this time it occurred to him what was happening—he was drifting through a downwind turn with a crablike motion. He let the wind drift him around a large maple tree in the middle of the pasture, keeping his eye trained on a piece of string tied to the front elevator. It blew straight back—proof that he was neither slipping nor skidding. He finally landed after covering 4,080 feet in 1 minute 35 seconds, the first man to really navigate the air.

Low-altitude training maneuvers today are usually flown at

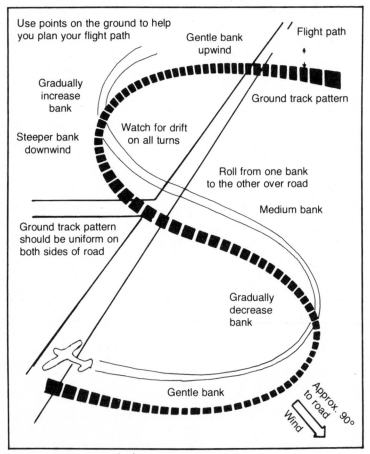

Fig. 7-2. S-turns across a road. (courtesy USAF)

about 600 feet, high enough to be safe and low enough for drift and altitude changes to be readily discernible. Flying through a full circle in a brisk wind, as Wilbur Wright did, quickly shows you why you appear to be slipping toward the ground turning downwind, and skidding turning upwind (Fig. 7-1). By cross-checking your slip-skid indicator, however, you see that it is just illusion.

One of the most elementary practice maneuvers in low flying is to make S-turns along a road running at right angles to the wind, so that each half of the S is equal, like two half circles. Here you are confronted with a whole new set of problems—flying with coordination, holding your altitude constant, and constantly varying the bank of your wings to compensate for wind drift (Fig. 7-2).

Once you can do good S-turns, go back to flying through a full circle, this time maintaining a constant radius around the center. It sounds downright simple, but at a recent flight instructor seminar I attended, more than half of the CFIs couldn't correctly decide that the steepest bank occurs in flying *downwind* (not crosswind, as most guessed)!

Once you become proficient in this maneuver, called *turns about a point*, you're ready to combine two such maneuvers, rolling from one into the other, this one called *elementary eight*. (Fig. 7-3).

A more advanced version of this maneuver is called the *pylon eight* or *eights on pylon*, and it is perhaps the most difficult of all low-altitude training maneuvers (not required for private pilot certification). In doing a good pylon eight, a pilot knows confidently that he can fly his airplane where he wants it to go while keeping his eye on the ground and not on the horizon line (Fig. 7-4).

The trick in flying a good pylon eight is finding your pivotal altitude, one in which a reference line from your eye level, paralleling the aircraft's lateral axis, is pinned on your ground "pylon" and stays there regardless of wind drift (Fig. 7-5). If you're too low, your turn takes you ahead of your pylon; if too high, the pylon moves ahead of your reference line.

Hence, to do good on-pylons, you must vary your altitude slightly to correct for groundspeed changes due to the wind. Here, as in elementary eights, you select two pylons (road intersections, trees or barns—but *not* cows; they don't hold still!) far enough apart to fly straight and level a few seconds between each one.

A master at pylon flying was the late Nate Saint, a missionary jungle pilot who developed a remarkable way to use the maneuver for air-to-ground communications in South America. The story is that Nate was called out one day to pick up a message from a medical

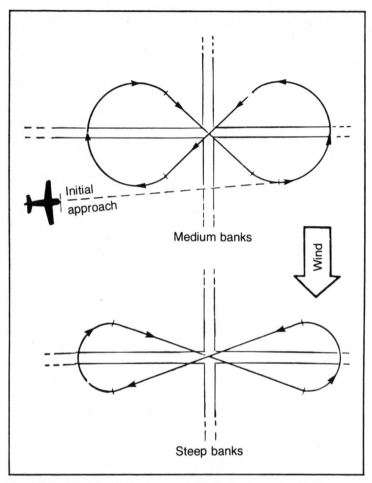

Fig. 7-3. Elementary eights across road intersection. (courtesy USAF)

expert on the ground in a clearing near a small jungle village stricken by an epidemic. The lives of the villagers depended on immediate diagnosis.

Saint found the clearing too small to land his lightplane in, and the medic had no radio. In desperation, Nate knocked the bottom out of an emergency water bucket and attached a long line to it, trailing it out behind as he flew in a circle at pivotal altitude. The wire spiraled down until the bucket hung motionless at the pivotal point, where the medic attached the message for Nate to haul in.

Saint radioed the medic's diagnosis to a distant rescue base, from which immediate medication was dispatched. The USAF

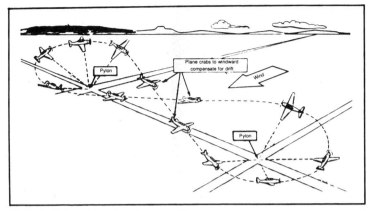

Fig. 7-4. Pylon eights. (courtesy USAF)

adopted what was known thereafter as "Nate Saint's Bucket Drop" for Arctic rescue work until the advent of helicopters made it unnecessary.

Another advanced flying maneuver useful in sharpening your coordination, orientation, and skill in power control is the precision 720-degree power turn, in which you zap around through a steep 360-degree turn in one direction and smoothly roll into one the other way. These should be performed with a bank between 45 degrees and the limit of your aircraft's performance. When you get them down pat, your CFI will personally buy you a beer for not spoiling his day.

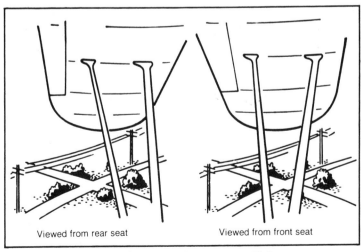

Viewed from rear seat   Viewed from front seat

Fig. 7-5. Reference points for pylon eights. (courtesy FAA)

Emergency Landings

Steep spirals—very, very useful in making emergency landings—are a sort of power-off version of the steep precision turn that will help you remain oriented under pressure and overcome any tendency toward vertigo. Your goal is to maintain a constant airspeed of about 1.4 times normal glide speed and a constant bank of around 60 degrees through 1080 degrees of turn (three full circles), recovering within 10 degrees of your entry heading.

In application to emergency landings, the steep spiral is made directly over the point of intended landing, correcting for wind drift and recovering near 1,500 feet to enter a standard 360-degree overhead approach and landing.

In actual forced landings, it's the last few hundred feet that fools pilots. Fearful of undershooting, they are wont to crowd close to the edge of the field until they find themselves in an impossible position to glide in safely and so overshoot. For this reason it is wise to establish a "base leg" sufficiently downwind to make a normal glide approach to land when you have dissipated enough altitude.

It is important in establishing your base leg for a forced landing to always *turn toward your field* keeping it in sight, flying down the sky in a series of figure eights until you have decided it's time to turn in and land. This decision can be made anywhere along the base leg, at a point called the key position.

Determining the moment to turn in and land involves an exercise in judgment based on experience, knowledge of your aircraft's gliding angle, and a simple trick of visually determining where that angle of glide will take you in relation to the horizon line.

For instance, let's say your glide angle is 10 degrees. That means the terminus of your glide is just 10 degrees below the horizon line, and it will remain there throughout the glide. Hence, drop your line of vision 10 degrees below your horizon and that's where you're headed.

Once you know where your glide is taking you, the next problem in visual perception is to determine just when to flare out, and here many pilots disagree on where to look.

Says the FAA *Pilot Instruction Manual*: "The distance at which the vision is focused should be proportionate to the speed at which the airplane is traveling."

Advised Aerosafe Inc.'s *Flight Instructor Handbook*: "The student . . . should be reminded to look a considerable distance down the runway, as if he were driving an automobile at similar speed."

Both statements are essentially correct, yet neither tells you *why* you must look ahead of your glide terminus to judge altitude near the ground. Visual reference points vary with angle of glide (not airspeed), but are essentially standard if you know what you're looking for, which is a critical change in perspective based on a simple triangulation process that one-eyed Wiley Post, a fine flier, well understood. Do a few deep knee-bends at the end of the runway and you'll get the idea.

What you are seeing is an angular change between any two reference points below the horizon, more pronounced as you get closer to the ground. Considerable research has been done to explain to student pilots just what an experienced airman looks for, and it is customary to tell students to "watch the runway at about the nearest point ahead of the aircraft where the ground isn't blurred."

Essentially you are looking for a sudden break in perspective, where the ground seems to tilt upward sharply, to find where to begin holding off and break the glide. Do this too soon and you may find yourself dropping in from a landing 10 feet too high; wait too long and you hit the ground and bounce.

Let's see what actually happens to the appearance of the ground where you are looking. Coming down a 10-degree glide path, your glide terminus *remains* exactly 10 degrees below the horizon, while more distant points move toward the horizon. A distant point 1,000 feet ahead of the terminus seems to drift horizonward at about one-half degree for each 10 feet of descent, hence the angular distance between the terminus and the point 1,000 feet further down the runway *expands* in your vision—until you reach a critical altitude where the process reverses.

The ground just beyond the glide terminus suddenly appears to recede. On a 10-degree glide, this occurs at an altitude between 30 and 40 feet if you're looking 50 feet beyond the terminus. (The closer to the terminus you look, the lower the altitude at which this critical prespective break appears.)

This is the visual cue you are seeking, the tip to begin holding off, and of course it will vary with the steepness of your glide path (the steeper the glide, the higher it appears).

As a rule-of-thumb, the flatter your glide angle, the farther away from the glide terminus you should look. Also, keep your eye moving, measuring the angular distance of distant reference points below the horizon, until you see the perspective "break." It takes a little practice, but it's easier when you know what you're looking for.

Carburetor Ice

This might be a good time for a discussion of one consistent cause of accidents due to engine failure taking off from or landing at airports—carburetor icing (Fig. 7-6).

I remember a warm, cloudless summer day at San Diego's Lindbergh Field when I sat in number two takeoff position behind a sharp little Monocoupe, waiting for the green light with engine idling. We'd been sitting there for a number of minutes, waiting for half a dozen aircraft to land.

The Monocoupe's pilot, a friend of mine, was a careful fellow who had preflighted his ship well to make sure the last leg of an excursion flight home from Ensenda to Burbank would be uneventful. There was nothing to indicate that two minutes later both he and his fiance, riding with him, would be dead.

The Monocoupe finally got a green light, rolled out into takeoff

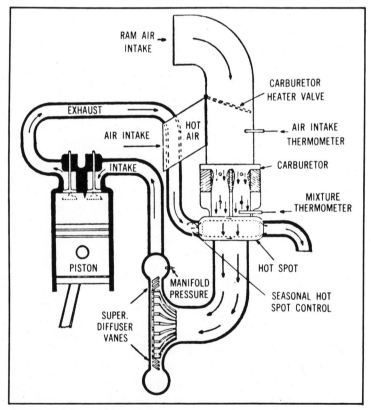

Fig. 7-6. Carburetor icing normally occurs at the intake throat. (courtesy FAA)

position, gathered speed, and then began to climb easily into a 10-knot headwind to perhaps 500 feet. Suddenly the left wing snapped down into a vicious spin. There was an explosion and a ball of fire as the hot little aerobatic ship burst into flames.

That happened several years ago. Recently, out of curiosity, I checked back through the accident reports to find out what the investigators had learned. The Monocoupe, a hot little ship, had gone into the spin when the pilot tried to turn back and land downwind after his engine failed. But why did the engine quit? Both tanks were full, and as the pilot had waited several minutes to take off, the assumption that the fuel selector valve was turned on was justified. Seeking other probable causes, the investigators concluded that careburetor icing was to blame.

This came as something of a shock. How could that have happened on a warm summer day with the temperatures in the 80s. I learned that it wasn't the heat, it was the humidity that was to blame—that and the fact that the pilot was presumed to have sat with engine idling *without applying careburetor heat.*

Contrary to popular belief, aircraft engines are more subject to carburetor icing on warm, moist days of early and late summer than in the wintertime. When free air temperature ranges between 45 and 85 degrees and humidity is high, *be careful!* Just as your home refrigerator may need defrosting more often in summer than in winter, so it is with your aircraft engine.

In winter flying, when free air temperatures on the ground range from 40 degrees Fahrenheit downward, insufficient moisture is generally present to cause carburetor ice to form.

Danger comes when fuel vaporization withdraws heat from the intake air, reducing it as much as 60 degrees as it passes through the carburetor throat. Thus free air at 80 degrees F. is chilled to 20 degrees F., well below freezing. If moisture is present, it precipitates as ice.

Sitting and waiting for takeoff with engine idling, air entering the carburetor may not be heated sufficiently to prevent an ice buildup, now suspected as the cause of many unexplained takeoff crashes.

A good number of pilots feel that consistent use of careburetor heat when atmospheric conditions are not conductive to icing is unwarranted, wasteful of fuel, and possibly harmful to your engine. Preheated air is less dense than free air and can cause an overly rich fuel-air mixture, loss of power, and perhaps engine detonation.

On the other hand, judicious use of carburetor heat can be a life-saver; it is well to remember that the most dangerous time for icing to occur is on takeoff or on a go-around. Pilots who do not consider themselves expendable make a practice of using careburetor heat while awaiting takeoff clearance as well as during landing approaches with engine idling.

Under early and late summer flying conditions, many pilots make it a rule to apply carburetor heat a full minute before reducing power to begin their approach to land, thus clearing the carburetor throat of any unsuspected icing.

How do you know when ice is forming? There are four good clues to remember: A drop in rpm with fixed-pitch propeller craft; a drop in manifold pressure in constant-speed propeller craft; roughness in engine operation; and backfiring.

The cure, of course, is to apply carburetor heat at once while seeking a spot for an emergency landing if necessary. If this doesn't clear things up, try leaning out your mixture to induce backfiring, which sometimes can jar carburetor ice loose.

Remember, however, that use of carburetor heat costs you a one percent loss for each 10 degrees F. temperature rise, so follow your manufacturer's recommendations, especially when you don't have a carburetor heat gauge.

The Private and Commercial Tickets

The main difference between a private pilot and a commercial pilot, as we said earlier, is that the latter can fly for hire. This does not necessarily mean he is the more knowledgable, for the private is the peer of the commercial pilot in many respects.

True, the commercial pilot must have logged 200 hours of flight time and passed more stringent oral, written, and flight tests, but there are many private pilots flying today with knowledge and skill equal if not superior to that of some commercial pilots.

Generally speaking, however, when a pilot becomes qualified for his commercial ticket he will go ahead and take the required tests, unless his interest in flying is confined only to flying his own airplane for pleasure or business.

In 1964, the old Civil Air Regulations were amended to preserve the traditional right of private pilots to share expenses with passengers, or to operate an aircraft for compensation or hire in connection with any business or employment if the flight is only incidental to it.

In other ways the added privileges of the private ticket have meant added responsibilities, as in passing a flight check to demonstrate ability to fly by instruments. This requirement by no means qualifies the private-only pilot to conduct a flight under an IFR (instrument flight rules) flight plan, but it does show the FAA that he has basic knowledge and skill in reading his instruments and safely controlling his aircraft by referring to them alone. The Private instrument flight check covers six items:

1. Recovery from the start of a power-on stall.

2. Recovery from the approach to a climbing stall.

3. Normal turns of at least 180° left and right to within plus or minus 20° of a preselected heading.

4. Shallow climbing turns to a predetermined altitude.

5. Shallow descending turns at reduced power to a predetermined altitude.

6. Straight and level flight.

If the pilot is also after a multi-engine rating, he will be expected to demonstrate skill in handling his aircraft with one engine shut down. He will also be flight-tested on a cross-country run covering four salient points:

1. Cross-country flight planning.

2. Cross-country flying.

3. Cross-country emergencies (lost, weather, overheating engine, power failure, etc.)

4. Use of radio aids to VFR navigation.

In the next chapter we will plan a typical cross-country flight and discuss the differences between "airport flying" and really getting out and going someplace. For after all, that's what an aircraft is for—to go somewhere!

Chapter 8
Cross-Country Flying

Your first trip away from the local airport, with your instructor beside you, is a revelation. Suddenly your airplane is more than just a means of getting off the ground. It is a machine for swift, efficient transportation, for getting from A to B, flying free as a bird far above crowded surface expressways.

A cross-country flight begins with flight planning, an art no more difficult than getting out of the old Rand McNally and planning an automobile excursion out Interstate 40 to Grandmother's Roadhouse, or a weekend trip to the mountains pulling a loaded trailer.

Charts

Instead of a Rand McNally, you will be using a more sophisticated map called a *Sectional Aeronautical Chart* on which geographical features, cities, towns and hamlets, roads, railroads, and even wagon trails are shown. Also included are radio navigation systems and a world of other useful data—virtually all you need, with the symbols explained on one foldout panel.

There are other aeronautical charts of different scales for faster aircraft, for jet navigation planning, and for local area flying, but let's stick with the Sectional right now.

This remarkable map is quite a bargain, containing a wealth of information, and is available from the Coast and Geodetic Survey, a branch of the Department of Commerce's Environmental Science

Services Administration (ESSA). It is a revision of earlier Sectional charts, designed to meet the requirements of three government agencies—the Departments of Defense and Commerce, and the FAA.

Sectionals, as they are called, are scaled 1:500,000, or eight statute miles to the inch, and as they actually contain less data than their predecessors (to avoid confusion), the FAA recommends that you carry along an *Airman's Information Manual* (AIM) to find information not on the map.

The new charts, which began appearing in December 1967, list only primary civil control tower radio frequencies in boxes of type near the airports, with secondary frequencies listed in the margin.

Other features of the Sectionals are a contour interval of 500 feet (to satisfy DOD requirements) and printed frequencies of Flight Service Stations for general aviation use. Measuring roughly 55 by 21 inches and printed on both sides, Sectionals are difficult to work with in the cockpit until you get used to folding them properly, and this is just one reason why flight planning is best done on the ground, prior to takeoff (Figs. 8-1, 8-2).

There are many other navigational charts available. The Coast and Geodetic Local Aeronautical Charts are scaled at four miles per inch and are excellent for use on VFR flights into congested areas, giving more topographical detail than Sectionals. C & G's Aeronautical Planning Charts, scaled at 80 miles per inch, are useful for planning long-range flights.

Fig. 8-1. Sectional chart is cumbersome to work with, so do it outside the cockpit.

Fig. 8-2. Chart is now folded conveniently for use in cockpit.

The Sectionals contain a combination of both geographical and cultural features (such as mountains, rivers, cities, and roads) overlaid with blue-tinted route information on the FAA's system of VOR aerial highways. This enables a pilot to navigate by radio and crosscheck his position by pilotage (flying by landmarks). Many pilots today depend to a large extent on radio aids for cross-country flying and so miss much of the enjoyment of flying by pilotage, an art that can come in handy should the radio suddenly go out.

Another type of cross-country navigation is flying by *dead reckoning*, a phrase which is a bastardization of *d'ed reckoning*, a contraction of *deduced reckoning*. It goes back to early days of maritime navigation, when a sailor knew that if he sailed east for 24 hours at 5 knots (nautical miles) he would find himself 24 × 5 or 120 nautical miles east of where he started. Of course, wind and tide affected his course, just as movement of the air affects the course of an aircraft.

Preflight Planning

As a first step in flight planning, you will, of course, decide where you're going, and by what route. The shortest distance—a straight line—may not be possible. (On long flights a "straight line" is *not* necessarily the shortest route; one follows a *rhumb line*, crossing all meridians at the same angle.)

There are also geographical reasons why it may be best to detour—to miss a high mountain, say, or to stay close to available emergency landing areas. Your Sectional chart also lists Prohibited, Restricted, Warning and Caution areas to avoid. (You *can* cross a caution area—by using caution: permission must be obtained to overfly the other areas.)

Preflight planning must also include a check on the latest weather reports and forecasts, consideration of fuel requirements, and provision for an alternate route if bad weather develops en route. You will also get together a few necessary items—a computer, plotter, and pencil to work out navigation problems en route; a flashlight for night flights; a supply of water and emergency rations for flights over remote desert country.

A personal visit to the nearest FSS (Flight Service station) is the quickest and best way to get a briefing on both weather and other en route hazard data. If you telephone the FSS or the Weather Bureau office, identify yourself as a pilot and state your intended route, destination, intended time of takeoff, and approximate time en route, and whether or not you will be flying strictly VFR.

If you find that the weather ahead poses no danger, get busy with your Sectional chart and draw a course line from your airport of departure to your destination. This may include a few doglegs, as in flying from VOR to VOR, or checkpoint to checkpoint to miss a mountain.

Study your chart for good en route landmarks such as cities and

Fig. 8-3. Sandstone peak near Monument Valley makes good landmark.

towns, lakes and rivers, airports and railroads, and be sure you aren't flying through Restricted zones (Figs. 8-3, 8-4).

Check carefully the terrain elevation. If you're planning to fly at 3,000 feet or higher, you'll have to conform to cruising altitudes appropriate to your direction of flight.

Fig. 8-4. You can't miss Meteor Crater for a landmark!

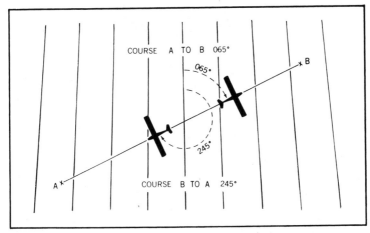

Fig. 8-5. Courses are determined by reference to meridians on aeronautical charts. (courtesy FAA)

Make a list of navigation aids along the route, whether you plan to use them or not. Even when flying by simple pilotage or dead reckoning, it's sensible to use your radio for cross-checking when possible. List the facilities that have voice broadcasts where weather reports are available en route. List also the Flight Service Stations along your route, and which frequencies are active for both receiving and transmitting. Other pertinent data is available in the *Airman's Information Manual*, but a check with the nearest FSS will normally give you all the information you need for a safe trip.

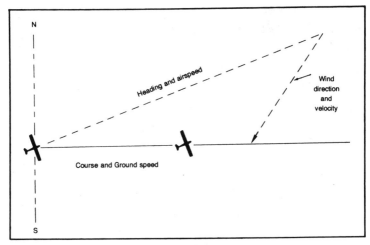

Fig. 8-6. Basic wind triangle. (courtesy FAA)

If you've paid attention to your ground school instructor, you'll know how to plot your course properly, correcting for magnetic variation and constructing a simple "wind triangle" to decide which magnetic compass heading to use to get you where you want to go. We won't go into all these details here, except to suggest that you brief yourself as fully as possible before starting out, thus taking the guesswork out of your flying (Figs. 8-5, 8-6).

Dead Reckoning

To fly a cross-country proficiently by dead reckoning, a good knowledge of how a magnetic compass works is essential. Following compass courses and not landmarks is the key to this type of navigation. It's very easy to mistake one mountain peak for another (Fig. 8-7), or even one town for another—particularly in the prairie country, where vast fields of waving grain stretch for miles in every direction, bounded by look-alike roads that lead to communities which are difficult to tell apart from the air.

I remember the sad plight of a World War II buddy, Big Dog MacDonald, who became confused on a training flight from a pre-glider school at Pittsburg, Kansas. After a half-hour of spin practice above a broken stratocumulus layer, Big Dog glided down and headed for what he thought was the Pittsburg airport.

On closer inspection, he found that there was no field at the northwest corner of the town! The church steeple was there, the bus station, and the grain silo, all as he had remembered them. He

Fig. 8-7. Sawtooth Range mountain peaks all look alike.

had the awful feeling that somebody had stolen the airport! In desperation he flew off to look over another crossroads town a few miles away—no airport.

Dusk began settling, and the stiff westerly wind that had blown him away from his practice area subsided a bit. He tried one more town, and once more his heart sank. Finally Big Dog realized he needed assistance, but he had no radio to make a Mayday call, and radar was then in its infancy, in use only in combat zones. A brilliant idea suddenly struck him—he whipped off his flight cap and wrote HELP! on the brim, then threw it out the window!

Had Big Dog been further along in his flight training and known how to use a magnetic compass, chances are he wouldn't have let the wind trick him. (He eventually landed, in a small field, and phoned his C/O to come find him.)

Using the magnetic compass, a pilot can combine dead reckoning and pilotage, navigating by calculations based on airspeed, compass heading, wind direction and velocity, and elapsed time, and constantly correcting for wind shift and other errors by checking landmarks.

In laying out a dead reckoning problem, you start with your *true course* line from Airport A to Airport B, measuring its relation to true North with a protractor. Next, correct true course for magnetic variation in the local area (shown on your Sectional as an isogonic line of red dashes), adding westerly variation or subtracting easterly variation to arrive at your *compass course*. (Some pilots mumble: "East is Least, West is Best, going from True to Compass.") Then check your magnetic compass deviation card for any instrument error, adding or subtracting as required to find your *compass course*, the heading you steer. To remember this sequence, our master of mnemonics repeats: "TVMDC—True Virgins Make Dull Company."

A magnetic compass, although the *only* completely independent direction finder in your panel, has its own limitations. In rough air it will swing back and forth in sometimes wild oscillations; to try to follow it is folly. Instead, average out the swinging and aim down the middle.

Controlled by lines of force of the earth's magnetic field, a magnetic compass is so balanced on its pivot that it will *tilt* if you bank your airplane while heading north, and tilt the opposite way when headed south. It also is affected by acceleration forces that cause it to swing in a climb or dive, accelerating or decelerating on an east or west heading (Figs. 8-8, 8-9).

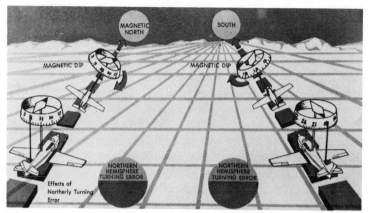

Fig. 8-8. Northerly turning error produced by magnetic dip. (courtesy FAA)

Expert pilots can make perfect instrument turns with a magnetic compass alone by knowing just how much to overshoot or undershoot in rolling out, but for the beginner, a simple rule is to fly straight and level for 30 seconds to let the compass settle down, *then* take a reading.

Some pilots are wont to overlook compass deviation errors in plotting courses, because they are usually so small (one or two degrees), but there are times when deviation can be catastrophic, as in the case of one pilot who flew out to sea and vanished on a routine night flight up the Atlantic Seaboard from Charlotte, North Carolina, toward Washington, D.C. Accident investigators learned that the pilot had made a gas stop at Richmond, Virginia, and interviewed the man at the gas pit to find out if he remembered anything abnormal.

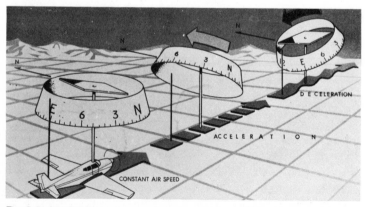

Fig. 8-9. Acceleration error upsets mag compass in flight. (courtesy FAA)

"Well, yes," the pit man recalled. "He had this big flashlight on his panel, right next to the compass, and after he used it to check the gas tank visually, he put it right back there."

There was the answer: He had followed his flashlight to his death; its magnetic clamp had thrown his compass off perhaps 90 degrees! The moral is: Keep all metallic objects such as flashlights, knives, and *particularly* magnets well away from your magnetic compass.

Pilot orientation is a key to good airmanship; you should *always* know where you are. It's not enough just to say "I'm somewhere on this here line from A to B" because emergencies have a way of happening when you least expect them. If you know your precise position, you can quickly work out an alternate airport problem without first having to determine where in blazes you are. By drawing a new course line from your position to the nearest alternate field, you can estimate heading, distance and the time to get there, and if need be, place an emergency call to the nearest FSS. It is this ability to improvise in a hurry and compensate for incomplete or inaccurate information that makes a good pilot.

In actual practice, in lower-level cross-country flying at cruising speeds in the neighborhood of 100 mph, you are likely to find that weather forecasts are not precise. The atmosphere is an ever-shifting sea of air and 100 percent accuracy in forecasting is not yet possible; even if en route conditions are known accurately now, they will not be the same six hours from now.

For this reason, experienced pilots learn to make rough estimates of wind conditions that are highly adequate for visual flight, correcting them as the actual flight progresses.

One method of making en route corrections is quite simple. Make a mark on your Sectional chart every ten miles along your course line, and with your computer check your actual ground speed against estimated ground speed as you progress along your route. Say your true air speed is 100 mph; in ten miles you will have flown six minutes (one tenth of an hour) under no-wind conditions. If you cover five 10-mile marks (50 miles) in 40 minutes, your computer shows that you're flying over the ground at only 75 mph and you're bucking a 25-mph headwind.

Similarly, if at the end of a half-hour of flying at 100 mph true airspeed, you find you're off course two miles to the right, you have a four-mph wind drift component from your left. You will see by checking with your protractor that you're off course about 2½ degrees to the right. To get back on track, double the error and (in

Fig. 8-10. Bend in Mississippi River makes good landmark.

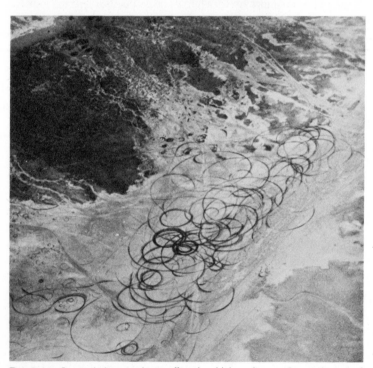

Fig. 8-11. Scarred desert where offroad vehicles play marks north end of Panamint Valley in California.

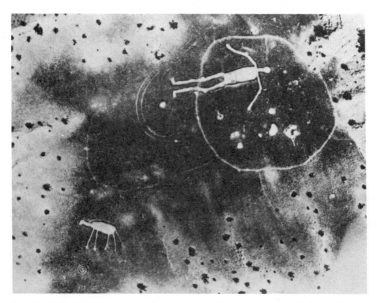

Fig. 8-12. Prehistoric archeological figures on desert mesa near Colorado River.

this case) turn left 5 degrees, then continue on your way with a 2½ degree left correction—until the wind shifts again!

One of the fascinating things about flying by pilotage is picking out landmarks and seeing if you can find them on your chart and vice versa (Figs. 8-10 through 8-12). For the beginner, estimating distances from air may appear difficult, but a little practice quickly orients you to distance judgment by "eyeballing."

A good practice is to compare your altitude above terrain with an equal distance to either side of your track, by looking down at a 45-degree angle. If you're a mile high, that barn over there, 45 degrees below the horizon, is a mile away.

Another good habit to form is estimating the height of mountains along your route to help identify them as distant landmarks (Fig. 8-13). Remember that for all practical purposes your horizon line remains the same, whatever altitude you fly. If you're cruising at 5,000 feet and that snow-capped peak on your right sticks up above the horizon, look for a mountain peak on your chart more than a mile high. If the peak is below the horizon, it's less than 5,000 feet high.

In the above instance we are referring to altitudes above mean sea level, not above the average terrain. On all but local airport flights, it is essential that you set your altimeter to local field

121

elevation before takeoff and, en route, check it against weather reports at Flight Service Stations or other navaids with voice facilities. On request, you will be given local altimeter settings in terms of barometric pressure converted to standard sea level conditions (29.92 inches of mercury is the mean figure).

Density Altitude

If your destination is a mountain airport, your concern is the local *density altitude*, which is not a measurement of height as are true altitude, pressure altitude, indicated altitude, or absolute altitude. It is the current condition at the airport station, and is affected in order by temperature, station pressure, and humidity (Fig. 8-14).

In the heat of the day, density altitude at a mile-high airport may be 7,000 feet; taking off in this thin air is far different from operating at sea-level fields. An expert on mountain flying conditions, the U.S. Weather Bureau's Tom Beecroft, meteorologist at Sacramento Airport, gives this bit of good advice:

"Many lower altitude fliers take their aircraft's performance for granted. Air is abundant at lower levels. Operations at the higher altitudes under high temperatures makes horsepower look like pony-power. The prop doesn't seem to bite the air as it should. Lift of the wings seems inadequate. The carburetor gasps for air. Mountains seem to grow faster than you can climb.

Fig. 8-13. Estimating height of Rocky Mountain peaks helps in pilotage.

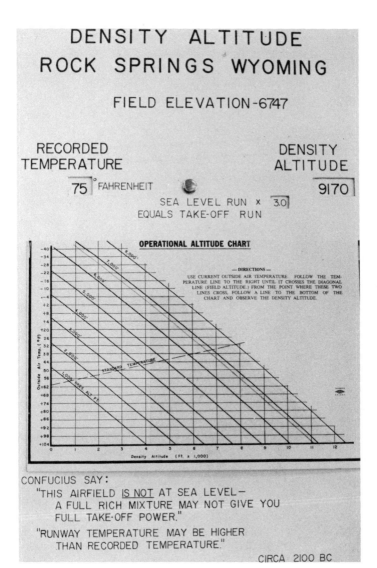

Fig. 8-14. Density Altitude (DENALT) Chart at Rock Springs, Wyoming, Flight Service Station contains good advice.

"Let's cite a hypothetical example: You fly into a high mountain airport just before lunch. The air is cool and refreshing. By mid-afternoon you decide to take a passenger up for a joyride to see the sights. The temperature is in the high 80s. Your passenger weighs 105 lbs. At the airport you look at the density altitude chart with average takeoff runs shown. After a quick check with the Aircraft Operations Manual performance chart, you note that your little passenger now weighs an effective 200 pounds. The joyride is off!"

Incidentally, you can buy a gadget called the Denalt Performance Computer from the Superintendent of Documents, Governent Printing Office, Washington, D.C., 20402. From it, a wide range of takeoff distance and rate-of-climb figures can be extracted for use under varying density altitude conditions. This computer replaces the old Koch Chart which appeared on the back of earlier Sectionals with hypothetical data of little or no practical use to pilots.

Why Pilotage?

In the next chapter we will discuss in greater detail the art of radio navigation, today one of the simplest ways of flying cross-country. In closing this chat on pilotage and dead reckoning, let it be said that the best pilots maintain their proficiency in these techniques, not only for the enjoyment of it but for two other valid reasons—it is good to fall back on in the event of radio failure, and on many cross-country flights you will be flying off-airways, with no navaids to guide you.

Then is when the world is your map . . . the stretching valleys and hills, the meandering rivers that which you can follow beyond the horizons in the way early pilots who carried the airmail followed railroad "iron beams."

There are many fine manuals to study and so become proficient in this art; to excel in it you will have mastered many of the demands of good airmanship. You will know without having to stop and calculate that you can make a round trip faster under no-wind conditions than you can when a wind is blowing (you'll fly longer in a headwind condition than in a tailwind condition and so increase your total time). You will know that your craft does not weathercock flying in a crosswind, the way it wants to on the ground. And most important, you will know that you don't need a radio in your airplane to fly cross-country under a VFR flight plan!

Flight plans are discretionary (except under certain circumstances) and the time and effort it takes to file one is small compared to the value received in knowing that the Fatherly Aviation Administration is keeping a watchful eye over you to ensure your safe arrival.

However, let's say you're ferrying a Piper Cub from Los Angeles to Phoenix with no radio equipment aboard. As long as you follow the "hemispheric" rule and fly at proper altitudes and remain VFR, you're clean. But you'd be cheating yourself—radios are important for emergency contact as well as normal NAV/COMM talk, and best of all, for obtaining en route weather reports on what's ahead.

You can call up the En Route Flight Advisory Service (Flight Watch) on 122.0, give your position and ID, ask whether it's snowing or blowing at your destination, and they'll be downright happy to oblige. In return for this fine service, you can do a favor for them and other pilots by making a *PIREP*, a pilot report on en route weather conditions you've encountered—cloud tops or bases, cloud layers, visibility, turbulence, haze, ice, thunderstorm activity—or maybe a UFO, if you happen to spot one zooming past.

As a responsible citizen of the airman's world, you thus become an active member of a fraternity of people who share the sky together, and who in emergencies will bend every effort to help each other.

In the past, incidentally, one of the traditional emergency procedures to follow when lost above the clouds (or inside them) was to fly triangular patterns when radio contact failed. If you can hear but not transmit, do this: Fly one side of an equilateral triangle by holding a compass heading for two minutes, then make a right turn at a rate of 1½ degrees per second through 120 degrees. Roll out and fly your second two-minute leg, make a second 120-degree turn, and fly the third two-minute leg. Complete a minimum of two such patterns before resuming your original course, and repeat at 20-minute intervals.

If your radio is completely dead, fly the same triangular pattern, except to the left instead of to the right (Fig. 8-15).

The presumption is that a surveillance radar observer on the ground will see you tracing triangles on his scope and initiate rescue action. If your receiver *is* working, tune it to the emergency frequency of 121.5 mc and follow instructions. If your radio is dead, the radar man will try to guide a rescue ship to intercept you so that you can follow him down to safety.

In actual practice, the FAA has found that busy radar operators are not likely to spot your pattern under the pressure of controlling IFR traffic, hence a new procedure recently has been established. General aviation pilots are asked to carry metallic chaff (strips of aluminum foil in bundles) and in an emergency toss it out at regular two-mile intervals. (This is the same kind of chaff used in World War II to jam enemy radar.)

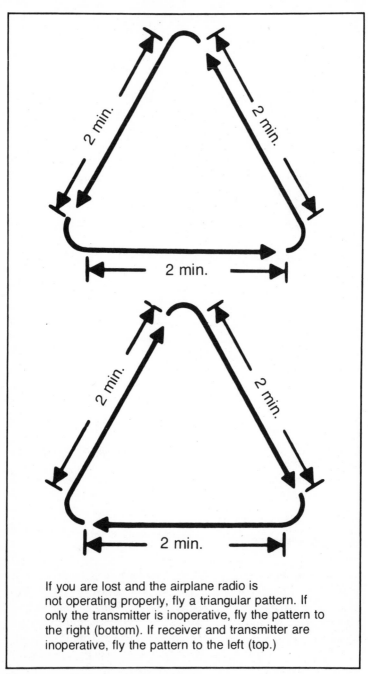

If you are lost and the airplane radio is not operating properly, fly a triangular pattern. If only the transmitter is inoperative, fly the pattern to the right (bottom). If receiver and transmitter are inoperative, fly the pattern to the left (top.)

Fig. 8-15. Triangle pattern to fly when lost.

These and other emergency IFR and VFR procedures should be understood in advance of flying cross-country, although the most important one of all is to *stay out of weather*!

In case you inadvertently do wander into sky conditions where you can no longer fly by reference to the Earth's horizon line, or at least see the ground beneath you, there is a simple instrument technique to get you back where you belong without losing control of your aircraft in a graveyard spiral. The AOPA 360° Rating procedure, a safety project of the AOPA Foundation, Inc., is based on a manual prepared by the Ohio State University's School of Aviation.

Basically, it teaches you attitute instrument flying proficiency based on scanning, interpretation, and aircraft control, replacing the natural horizon with the artificial horizon on your instrument panel. And it should be mastered under guidance of an instrument-rate CFI. Thousands of student and private pilots who did not learn to fly under the integrated primary flight curriculum have benefited by this AOPA 360° Rating, which will be discussed in another chapter on instrument flying.

One of the dangers in cross-country flying, for the student pilot, is miscalculating time and getting caught out after dark. Night flying is not difficult, but there are a number of things to know that your CFI will explain before you make your first night solo.

In the first place, the three-dimensional world below appears two-dimensional after dark. The horizon and landmarks are obscure; the choice of emergency landing fields is limited.

Fig. 8-16. Freeway light patterns appear at left of approach lights to runway of airport at center.

Cockpit illumination must be carefully controlled to provide ample reading ability of all engine and flight instruments without disturbing your night vision. A small flashlight is handy in case of failure of instrument or cockpit lights. Navigation and rotating beacon lights must be checked prior to takeoff.

As discussed earlier, most pilots use some power on approaches when making night landings for more positive control. When nearing the flareout level, a distinct change in perspective of runway lights can actually make a night landing simpler than a day landing. In most cases, the night air is also more stable and heavier than in the daytime.

Flying cross-country at night, highways appear as strings of lighted beads, and small communities as clusters of fireflies (Fig. 8-16). There is much beauty at night—and danger too, particularly on a dark night with no moon, no horizon, and few surface lights for reference. Under such conditions disorientation or vertigo can distract the pilot, just as when flying in IFR weather inside a cloud. This alone is good reason for you to learn instrument-attitude flying early, a subject we'll come to later.

Chapter 9
Cloudland

In 1873 John Wise (Fig. 9-1), the dean of American aeronauts, sat down and wrote the story of his 40 years experience drifting through the solitude of cloudland, a region where today thousands of airmen routinely travel. To the average pilot weather is a menace, something to fear, to avoid. But to old John Wise there was beauty there: "You can see mountains and valleys, precipices and projections, lakes and rivers, domes and spires, and all the time the scene is changing. Sometimes beams of light are perceived issuing from above and illuminating these aerial castles with all the richest tints of refrangible drapery, and then again melting away into a mellow mist of dissolving vapor . . ."

In his frail balloons, Wise learned to ride the turbulent storm fronts with eyes blazing and whiskers blowing, enjoying to the hilt the black, ominous, rolling formations of cloudstuff sweeping across the land ahead of onrushing pressure waves (Fig. 9-2).

This pioneer aeronautical meterologist learned to read the language of cloudland—the portent of a mackerel sky of cirro cumulus; the threat of towering cumulonimbus, black, anvil-shaped thunderstorms laced with lightning and ripped with terrifying updrafts that lift hailstones big as oranges, and which today can wrench the wing off an airliner.

Wise loved to gaze down upon soft, fluffy blankets of stratus clouds and watch for graceful, slender lenticulars, arcing across the sky where the jet streams blow over mountain tops, creating

Fig. 9-1. Pioneer American aeronaut John Wise saw beauty in cloudland. He was first to discover the stratospheric jet stream, and planned to use it to fly to Europe in 1800s.

Fig. 9-2. John Wise saw weird cloudscapes from his balloons, including a startling mirage.

mountain waves in which soaring ships today can sail for hundreds of miles and reach dizzying altitudes above 60,000 feet.

Weather Facilities

From the days of this aerial poet until now, the study of aviation weather has grown from an art to a science, from guesswork to near-certainty. Out in space, Applications Technology Satellites hang like manmade moons, remote weather observation outposts in synchronous orbits 22,300 miles overhead—watching, photographing and returning to Earth instantaneous pictures of the worldwide weather pattern at a glance (Figs. 9-3, 9-4).

Balloons of thin Mylar called GHOSTS (Fig. 9-5) ride the stratospheric winds around the Southern Hemisphere, radioing back vital data on the mysterious mid-region of our atmospheric sea below satellite orbits and above the reaches of lower-floating rawinsondes. At a rocket launch pad at Thumba in Southern India, a U.N. program was launched to fire weather probes into equatorial skies to add to the total weather picture.

Fig. 9-3. Photographs of weather from space show complete storm systems like this one off the coast Morocco, North Africa. (courtesy NASA)

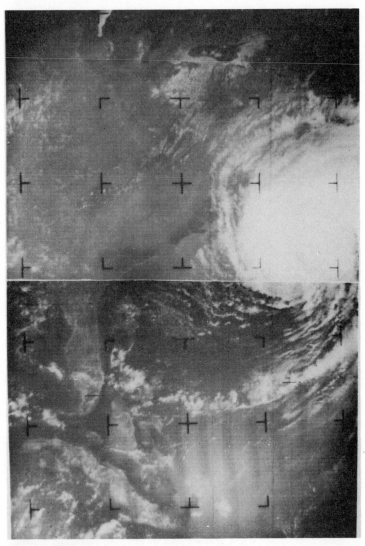

Fig. 9-4. Whirling hurricane photographed by NIMBUS II satellite off east coast of U.S. (courtesy NASA)

Another U.S. agency created in 1951, the World Meteorological Organization, added still more bits and pieces to the global weather picture; in the late 1960s a World Weather Watch was established to implement this agency and create a more effective understanding of total weather. To handle this growing mass of data inout, complex new computers in the U.S. Weather Bureau are busy at work creating new statistical models of the sky. Their goal is to eventually produce long range forecasts not only of tomorrow's weather, but for two weeks hence.

Until this gigantic effort becomes a workable day-to-day process, pilots must understand both the limitations and the capabilities of modern meteorology. It may well be more dangerous to place complete faith in weather forecasts than to have none at all.

Aviation forecasts are prepared by 51 National Weather Service Forecast Offices (WSFOs). Forecasts issued morning and midday are valid for 12 hours, those issued in the evenings for 18 hours. Pilots obtain the forecasts as FSS briefings, PATWAS (Pilot

Fig. 9-5. Mylar GHOST balloon (*Global HOrizontal Sounding Technique*) ride stratospheric winds to report on weather. (courtesy NCAR)

133

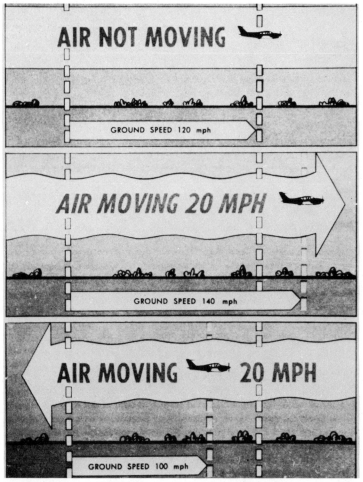

Fig. 9-6. Effect of winds on aircraft ground speed (airspeed is not affected). (courtesy FAA)

Automatic Telephone Weather Answering Service), TWEB (Transcribed Weather Broadcast), and Flight Watch stations on 122.0.

Air Route Traffic Control Centers are currently phasing in a new system of computer-generated digitized radar displays called Narrowband Radar, giving controllers a new, more precise look at severe weather that they can advise pilots to avoid.

Currently, forecasters average about 75 percent accuracy in predictions of:

- [ ] Passage of fast-moving cold fronts or squall lines within plus or minus two hours as much as ten hours in advance.
- [ ] Passage of warm fronts or slow-moving cold fronts within plus or minus five hours up to twelve hours in advance.
- [ ] Rapid lowering of ceiling below 1,000 feet in prewarm front conditions within plus or minus 200 feet and within plus or minus four hours.
- [ ] Onset of a thunderstorm one to two hours in advance if radar is available.
- [ ] The time rain or snow will begin to fall, within plus or minus five hours.

Studies at the Severe Local Storm Center in Kansas City, Missouri, show that one out of two tornado forecasts verifies, although flying in an area for which a tornado forecast has been issued is not necessarily hazardous. The probability of actually encountering a twister is "extremely low," says the Weather Bureau.

Because of the uncertainties of weather prediction, there is no better investment in your own safety than to study basic weather principles and learn to interpret and utilize the many fine services FAWS offers (Figs. 9-6 through 9-17).

On a cross-country flight in the spring of 1967, I was warned by FSS briefing at Monroe, Louisiana, that a "Tornado Alley" was expected to become active at 3 o'clock that same afternoon (it was then 1 p.m.).

The trouble zone was dead ahead, on a line running from Amarillo to Texarkana, where a Polar Cold frontal system was beginning to clash with a flow of Maritime air from the Gulf of Mexico. The warning was enough to call for a change of flight plan to fly further south by a good 100 miles—a wise decision, it turned out, because twisters *did* in fact strike Monroe that afternoon, while another sucked a houseboat from the river water near Shreveport!

This is why pilots prefer to have an alternate route and alternate airport in mind on a cross-country flight—unless they're flying an airliner equipped for Category II weather operations.

Weather can be the general aviation pilot's best friend, providing brisk tailwinds and azure skies, or his worst enemy, a gathering of elemental forces to avoid. Flying in his local area, the airman comes to know the vagaries of regional weather, but too often, traveling cross-country, he will encounter conditions he may not know how to deal with.

Flying along the Florida coastline, towering buildups of awe-

Fig. 9-7. Spectacular Sierra Wave over Owens Valley east of Sierra Nevada range produced violent roll cloud at left, sucking up sand off valley floor.

Fig. 9-8. High altocirrus clouds indicate direction of jet stream activity.

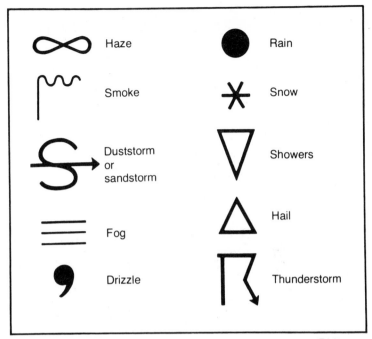

Fig. 9-9. Basic weather symbols used on weather maps. (courtesy FAA)

some cumulonimbus clouds are playthings to local pilots, one of whom told me, "We like to fly beneath them . . . it's cooler there!" A stranger to the sky over Florida, however, may not realize that powerful updrafts can sweep a lightplane into the seething mass of violence like a luckless moth.

At Daytona Beach, a radarman showed me on his scope what a line of thunderstorm cells looks like, building higher and higher in mid-afternoon. "Fly along the coast just offshore and you'll be okay," he smiled.

Sky cover	No clouds	1/10 or less	2/10 or 3/10	4/10	5/10	6/10	7/10 or 8/10	9/10 or over-cast with openings	Com-pletely covered	Sky obscured
Symbol	○	◔	◔	◑	◑	⊖	◕	●	●	⊗
Code	0	1	2	3	4	5	6	7	8	9
Term	Clear		Scattered				Broken		Overcast	Dust-storm, haze, smoke, etc.

Fig. 9-10. Weather map symbols for sky cover (courtesy FAA)

Fig. 9-11. Typical cumulonimbus "anvil" cloud formation.

The Central Plains region, the jousting ground of hot and cold air masses, is tornado country. Here a pleasant afternoon flight can bring you face to face with black funnels of death to be avoided!

Ground fog layers over the Mississippi Valley . . . supercooled air with built-in icing along the Great Lakes . . . wild turbulence in the lee of the Rockies . . . these are flight hazards to understand well and so live with, for they are no longer dangerous when you understand their limitations and fly accordingly.

Weather over the West Coast is directly affected by powerful energy changes taking place over the Pacific Ocean, through sea-air exchanges were only recently understood. One such weather phenomenon remained unknown to meteorologists until 1966 when spectacular color photos of the world's weather began coming back from the weather satellite ATS-1. This was an equatorial band of clear sky extending westward from South America to past the International Date Line.

One serious student of the sky, and a man who first understood what weather fronts were, is Professor Jakob Bjerknes, professor emeritus of meteorology at UCLA. According to him, the cloudless equatorial band is the result of submarine cooling caused by an upwelling of ocean water brought on by strong easterly winds that blow there at certain seasons. This equatorial cooling is the start of a chain reaction that ultimately determines weather patterns along the Pacific slope.

It is interesting to note that the western United States has been scoured by the same weather patterns for millions of years. At Grand Canyon National Park you can find dramatic evidence of what the weather picture was like 100,000,000 years ago, when the same winds we have today ranged northwest from the Gulf of Mexico to pile up shifting sand dunes where the Coconino Plateau now stands.

Fig. 9-12. Broken stratus deck covers Los Angeles streets.

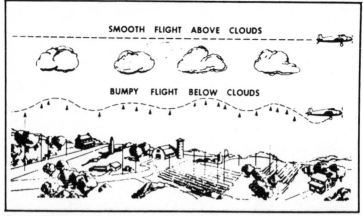

Fig. 9-13. To avoid convective turbulence, fly above cloud layer. (courtesy FAA)

Arid Arizona got little rainfall from the Pacific storms then. Today they are still wrung dry sweeping up the western slope of the Sierra Nevadas, where giant sequoias grow in lush rain forests. This same mountain range produces the familiar Sierra Wave on which soaring enthusiasts ride updrafts into the stratosphere.

Getting a Weather Briefing

Where do you go for a weather briefing? For a face-to-face chat with an expert forecaster, you simply visit the local Weather Bureau Airport Station (WBAS), the Weather Bureau's main link between the aviation community and processing centers. The WBAS weatherman has the advantage of knowing local conditions in addition to teletype and facsimile information from forecast centers, and so can discuss with you flight problems in practical terms.

At Flight Service Stations (FSS) of the Federal Aviation Administration, trained aviation weather briefers will help you inter-

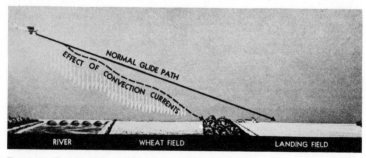

Fig. 9-14. Downdrafts over cool land, water affect glide path. (courtesy FAA)

Fig. 9-15. Air currents inside a cumulonimbus thunderhead. (courtesy FAA)

pret facsimile prognostic charts and teletype sequences offering weather advisories along your intended route.

In addition to personal visits to WBAS and FSS locations, there is up-to-the-minute aviation weather information as close as your bedside telephone via PATWAS (Pilot's Automatic Telephone

PRESSURE AT 5000 FEET = 25 INCHES

WHEN REDUCED TO SEA LEVEL— 25+5=30 INCHES

Fig. 9-16. Barometric pressure at weather station is reported as pressure at sea level. (courtesy FAA)

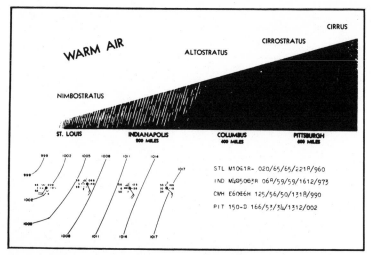

Fig. 9-17. Warm front overruns cold air, produces rain as it lifts. (courtesy FAA)

Weather Answering Service). The numbers to call are unlisted but are available to pilots in The AIM (*Airman's Information Manual*).

You can also tune in your aircraft radio (or a home receiver with appropriate frequencies) to obtain TWEB (continuous transcribed weather broadcasts) weather information covering a 250-mile radius, from many low and medium frequency FAA Navaids. In some regions, like Los Angeles and Kansas City, TWEBs also cover weather briefings for the major connecting airways over greater distances.

You can check your AIM for the local Weather Bureau number to call for person-to-person telephone briefings, obtainable as well from Flight Service Stations. Now under development commercially is a portable facsimile machine that can be connected to any telephone to obtain facsimile briefings. It will help give you the "big picture" via the National Weather Facsimile Network, which now extends to about 360 civilian and military weather offices.

What kind of weather briefing you need depends on the kind of flying you intend to do. An IFR-licensed pilot will be more concerned with ceiling and visibility data at his airports of departure, arrival, and alternate than elsewhere. His main interest is in cloud tops, weather hazards, winds aloft and at the surface, and clear air turbulence (CAT) reports.

The VFR pilot, on the other hand, will want a more complete briefing prior to departure. Important to him is whether cloud bases are high enough to fly below legally along his entire route, whether or not "VFR on top" flight is feasible, and whether visibilities are sufficient to navigate by pilotage.

In requesting a weather briefing, whether in person or by phone, remember that the briefer doesn't know you from a tomato picker unles you tell him. Give the man five bits of information and you'll do him and yourself a favor:

1. Tell him you're a pilot.
2. Tell him what kind of flying machine you will be operating (light single-engine, high-performance multi-engine, or jet).
3. Tell him where you're going.
4. Tell him when you plan to depart.
5. Tell him whether you can fly IFR.

With that to start on, the briefer then will give you the picture the way you want him to tell it, with a weather synopsis (positions of lows, highs, fronts, ridges, etc.), current weather conditions, forecast weather conditions, alternate routes to fly, hazardous weather, and forecast winds aloft. Like shopping in a department store, if you don't hear what you want, ask for it, but the weatherman normally will cover all the above points on his checklist.

In many FSS stations, the specialists may be pilots themselves; if you tell them you're a student pilot, chances are they'll be sympathetic and help interpret weather problems that may at first appear confusing. Simply ask the man which is the best route to fly, at what altitudes the best winds are, and where to look for trouble en route—building thunderstorms, frontal activity, CAT, etc.

Meteorology is a fascinating subject, and to attempt to cover all the things you'll need to know to be a good, weather-wise airman would take more space than we have here. Besides, a *little* knowledge can be dangerous! Until you learn to read the sky and know what the cloud patterns mean, listen to the expert—the weatherman—and remember that a lack of weather knowledge and failure to obtain a weather briefing account for a large percentage of General Aviation accidents.

In one recent year there were 4,400 such accidents reported to the FAA, of which 420 resulted in fatalities. Of the latter, 33 percent were attributed to loss of control of the aircraft by non-instrument rated pilots in adverse weather conditions. The biggest weather killers, incidentally, are not tornados or hurricanes, but simply low ceilings and below-margin visibilities.

To enjoy the skies, be weather-wise!

Chapter 10
On the Gauges

At 10 a.m. on the morning of September 24, 1929, Army Air Corps pilot Jimmy Doolittle climbed into the rear cockpit of a squat little Consolidated NY-2 Husky biplane and grinned up at an elderly man standing on the wing.

Harry Guggenheim, president of the Daniel Guggenheim Foundation for the Promotion of Aeronautics, slapped the pilot on the back, wished him good luck, and pulled a canvas hood over the cockpit.

Jimmy Doolittle was about to make history's first complete blind flight, from takeoff to landing. His only "eyes" were his instruments and a safety pilot, Lieutenant Benjamin Kelsey, who rode in the front seat, his hands outside the cockpit. He was there only for emergency takeover.

In a moment the Husky was thundering down the runway of Mitchel Field, Long Island, to begin a historic 15-mile flight to free airmen once and for all from dependency on the natural horizon. Jimmy lifted into the air smoothly and flew directly into the west for five miles, then gently turned 180 degrees to follow a shortwave radio beam back over Mitchel Field.

A small light glowed in his cockpit, activated by a radio marker beacon at the runway's end. Jimmy clicked his stopwatch, steadied his airspeed, and flew precisely two miles more to the east. He then turned 180 degrees again to begin a blind approach to the runway. A slight error in judgment would bring Lt. Kelsey's hands inside to grab the stick, but that did not become necessary. Doolittle expertly crossed the boundary at 50 feet, reduced power, raised the Husky's nose to kill off airspeed, and made a perfect wheel landing within feet of where he had started!

To achieve this remarkable milestone in aviation, Doolittle used a brand new instrument designed and built by Elmer Sperry, 70-year-old founder of the Sperry Gyroscope Company. Sperry had made a small fortune by adapting a child's toy to marine and aircraft uses; the instrument he built for Doolittle was a combination artificial horizon and directional gyro. He also developed for the experiment a visual radio direction finder with which to follow a radio beam.

Bernard Kollsman, a Brookly inventor, supplied a barometric altimeter more sensitive than any other yet designed, and for nearly a year Jimmy practiced takeoffs and landings with these pioneer instruments, many during the night and when fog rolled in over Mitchel Field.

Doolittle's achievement was duplicated on May 9, 1932, by Captain Albert F. Hegenberger, chief of the McCook Field Instrument Section, who made the world's first solo blind flight from Patterson Field. He received the 1934 Collier Trophy for his work, which included instrument flying with the aid of the new Kreusi radio compass, with which he could home on commercial broadcasting stations.

By 1936 Major Ira C. Eaker, another great pioneer pilot, flew an Army pursuit ship from New York to Los Angeles under the hood. Frequently he led his escort ship through zero-zero weather.

The Air Corps had been concerned with blind flying capability ever since the early days of the transcontinental air mail runs, when pilots had only flickering bonfires to guide them. Charles Lindbergh, during his career as an air mail pilot, twice bailed out of his ship and let it crash when lost above the clouds. There simply was no other safe way to get down!

One pioneer in blind flying who did much to convince the Army that instrument flight training was essential was Captain William C. Ocker. He had made the world's first long-distance blind flight, from Washington, D.C. to New Philadelphia, Ohio, relying primarily on the Sperry gyro horizon.

Captain Ocker introduced to the army flight program a training device called the Barany chair, in which a cadet was spun around and around in total darkness to prove to him that his senses were not to be trusted with no outside visual references to rely on; invariably

Fig. 10-1. Turn and bank indicators (2- and 4-minute turn types). (courtesy FAA)

the cadets believed they were spinning in the opposite direction!

In the early 1930s, others began turning their attention to developing a systematic approach to instrument flying. One researcher, Howard P. Stark, in 1932 evolved what was known as the 1-2-3 Method, in which the turn-and-bank indicator was the primary instrument. Students were taught to control the turn needle with the rudder, the ball with the ailerons, and the airspeed with the elevator (Figs. 10-1, 10-2).

Adopted by the Air Corps, the system was considered basic in that it did not rely on gyroscopic instruments (such as the artificial horizon), which too often could fail, particularly when icing knocked out vacuum-driven gyro gauges. It worked reasonably well, because basically the students coordinated stick and rudder anyway,

Fig. 10-2. Airspeed indicator becomes primary power instrument, altimeter primary pitch instrument flying straight and level. (courtesy FAA)

and so turn-and-bank readings were corrected accordingly.

The British Empire Training System in World War II applied typical English logic to the matter and reversed the system. Cadets were instructed to control the turn needle with the ailerons, because a turn is proportional to bank, and then trim the ball in the center with rudder pressure. Elevator-throttle coordination was also stressed in the RAF system, which provided cadets with integrated IFR/VFR training on every pre-solo dual flight.

By 1940 the value of using the artificial horizon as a primary blind flight instrument became obvious, with publication of a paper on the subject by Frederick Billings Lee of the Civil Aeronautics Authority.

Attitude Instrument Flying System

Today, with availability of more reliable gyro instruments, the FAA fully endorses the Attitude Instrument Flying System and recommends that all available instruments should be used (Figs. 10-3 through 10-5).

Most CFIs today have adopted the attitude system for instrument flight instruction, and for a guideline use an indoctrination course known as the *AOPA 360° Rating*. Student pilots who com-

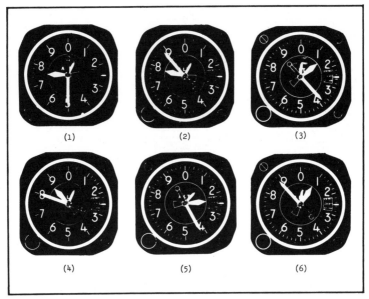

Fig. 10-3. Altimeter readings: (1) 7500′ (2) 7880′ (3) 1380′ (4) 8800′ (5) 12,420′ (6) 880′. (courtesy FAA)

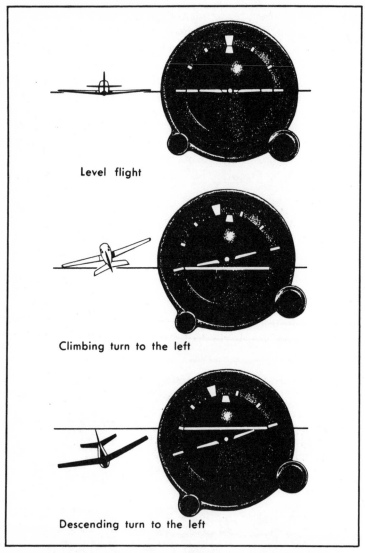

Level flight

Climbing turn to the left

Descending turn to the left

Fig. 10-4. Indications of gyro horizon (attitude indicator). (courtesy FAA)

plete this training may qualify for the FAA's Blue Seal certificate of proficiency and meet instrument flight test requirements for a Private Pilot Certificate.

Before discussing the atittude system in more detail, let's remember that the old 1-2-3 system is retained today under the name *partial panel flying*. The turn needle remains the primary instrument for determining wing position, though the angle of bank must be deduced from information available on rate of turn and airspeed, whereas with the artificial horizon the bank is pictorially represented.

In partial panel flying (the 1-2-3 system), the airspeed indicator becomes the primary indicator of nose position. Whereas with the artificial horizon you can see at a glance if the nose is too high or too low, because of lag in the airspeed reading without an artificial horizon you won't know this until too late. Like the rate-of-climb indicator, the airspeed can be thought of as a "trend" or "history" instrument.

The name 1-2-3 for this system of instrument flight comes from the sequence of corrective action used to maintain straight and level flight (or a properly banked turn). To fly straight and level, you first center the needle, then center the ball, to bring the wings level. The third step is to return the airspeed to normal cruise to correct for a climb or dive.

In both systems, a rapid eye shift is essential to good flying technique. Each time an instrument is checked, the eye returns to the primary instrument—the needle-and-ball in the 1-2-3 or partial panel system, the artificial horizon in the attitude system.

Inexperienced pilots will spend too much time attempting to correct only one flight condition, such as a too-high airspeed, letting the other conditions get worse. In the 1-2-3 system, a pilot may see the turn needle deflected to the left, the ball deflected right and the airspeed climbing above cruising. What is happening? He is in a sort of skidding spiral dive to the left. Correction: Center the needle

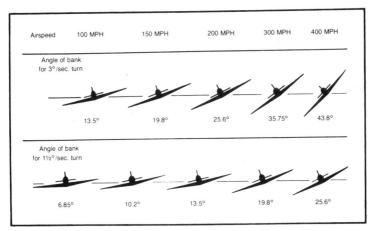

Fig. 10-5. Relationship of airspeed, bank, rate of turn. (courtesy FAA)

with the ailerons, thereby leveling the wings and stopping the turn, and at the same time kick the ball back to center with the right rudder to correct for skid. Then—almost simultaneously, in fact—apply a brief amount of back pressure on the elevators (the old speed control) to raise the nose and kill off excess airspeed.

One should *not* hold back pressure until the desired airspeed is reached because, remember, the airspeed indicator is a history instrument. By the time your needle returns to cruise speed, you've got your nose up too high! The result is a bad habit called chasing the airspeed, a never-ending game. Instead, make a small correction, then wait and see what the needle says a few seconds later.

Flying a "full panel" with the artificial horizon as your primary instrument is obviously, therefore, much simpler. Once your ship's attitude is right, with wings level and nose where you want it in relation to the horizon bar, there is only the power setting to worry about.

Vertigo

One of the greatest dangers in instrument flying is vertigo, a sensory disorientation in which either you or your environment seems to be whirling dizzily in utter denial of what your flight instruments are telling you. Some pilots will put their head down and shake it from side to side to change the sensory responses of the inner ear and reestablish a state of normal balance. Until the illusion passes it is quite essential to place full faith in the instruments, not in your own senses.

This vertigo problem was of course known to pilots in the 1920s, when the Barany chair was in wide use, and it works this way: Say you are flying straight and level, and inadvertently you let your right wing begin to drop—so slowly that the fluid in your inner ear is not affected by the tilt. Suddenly you see the turn needle deflected (or the gyro horizon tilted) and make a correction, bringing your right wing up sharply to level flight attitude. Your ear *does* sense *this* movement, and translates it as a tilt from level flight to a left bank. You then have the illusion that you are flying with the left wing down, when actually you are straight and level.

I experienced this once in a startling manner when climbing up through an overcast on an SID (standard instrument departure) from Lockheed Air Terminal in Burbank, California, on a magnetic heading of 120 degrees. The base of the stratocumulus layer was at 2,000 feet, and the tops were reported at 5,000. Actually they were higher, but as I passed 5,000 I began to see the clouds break up above me; patches of blue sky appeared.

I slid back the canopy of my BT-13 and began to fly VFR in a climbing turn to the right, staying out of the towering white cumulus clouds piled around me like cotton candy. In a moment I was back inside the clouds, still straining to see blue sky; instead, it seemed to grow darker. I glanced quickly back at my panel—the turn needle was far over to the left, the airspeed building!

For a second I *knew* the gauges were wrong! I had been spiraling upward to the right only a moment ago—I just could not be in a left spiral dive! Then I realized with a shock that my senses were lying to me. I had to force myself to fight them, to roll the ship to the right; it seemed like a suicidal thing to do, to roll inverted...

The turn needle slowly centered, and as I reduced power and killed off the diving airspeed, orientation returned. Had I not had instrument training, I might well have simply hauled back on the stick to fight the airspeed and wound up tight in a graveyard spiral, a non-habit-forming thing to do!

There are other freak sensations to watch for, many due to false visual stimuli. Once, making an intrument approach into Lockheed at the other end of a charter flight to Cedar City, Utah, I began a night letdown into an undercast heading due east, on the western leg of the old Burbank square range. Ahead and to the right of my nose was a full moon, a dull orange ball that sank into the softness of the cotton blanket. I was suddenly startled to see the same orange glow on my *left*! Had I inadvertently made a wild turn? Then I recognized the glow as that of my port running light, reflected from the vaporous cloud that engulfed me. Things immediately returned to normal.

These are simple illusions which can (and do) happen frequently to the average pilot. Others, far more dangerous, can occur if you let your ship wander into an unusual attitude, such as a steep climbing turn. Swift, sure recovery is simple when you know how, by using in coordination aileron, rudder, elevator, and throttle as required.

Once you have mastered the art of controlling your airplane's attitude and flight path by reference to instruments, you will sense an achievement that gives you a whole new confidence in flying. It is not a difficult thing to master, when you understand what is involved and learn what each instrument is for and how to blend their readings together into a whole new set of sensory responses. No longer will you listen to other false signals; visual perception alone will be working for you.

Now that you know how to control your ship in flight, the next step in practical instrument flying is to fly it in relation to radio checkpoints, the same way you applied your basic VFR flying skill to cross-country pilotage. America's vast network of electronic skyways is yours to venture onto in any kind of weather, once you are proficient at NAV/COM (Navigation/Communication) flying, VFR or IFR.

Chapter 11
Radio Navigation

Once you have graduated from small training aircraft to utility aircraft you will do most of your cross-country flying by radio navigation, actually the easiest and simplest way to fly today, whether IFR or VFR. While there are a number of different radio navigation facilities and systems in common use in the United States, the primary system is the Very High Frequency Omnirange System (VOR), which eliminates many of the drawbacks of navigating by radio compass or by aural signals of the vanishing low and medium frequency radio ranges. VOR information is presented visually, and the frequencies used (108-118 MHz) are free of static, on voice channels.

VOR

In a moment we'll file a VFR flight plan for a typical cross-country flight and see how the system works, but for the moment let's see what it is. Since the first VOR was commissioned in 1946, more than 950 stations form an interlocking network so spaced that you can either fly along the airways from one to another, or fly off the airways tangentally to the VOR radials.

Each VOR station provides 360 separate courses from (or to) its ground transmitter, one degree apart. These are called *radials*, and they extend like spokes of a wagon wheel from the station hub. You can fly along any one of these, either to or from the station or at an angle to it, and know exactly where you are.

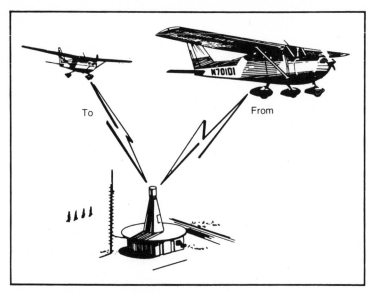

Fig. 11-1. VOR stations report whether you're flying toward or away from transmitter. (courtesy FAA)

More expensive aircraft radios are also equipped to handle DME (Distance Measuring Equipment) and TACAN (Tactical Navigation, military UHF equipment) broadcasts which give both distance and azimuth information to a pilot so equipped. For simplicity, we'll refer to VORTAC and VOR/DME stations as VORs.

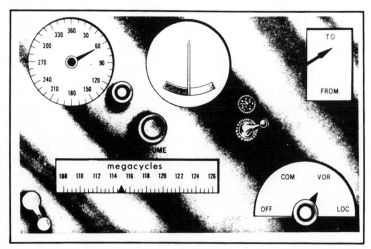

Fig. 11-2. Composite drawing of VOR omnireceiver. (courtesy FAA)

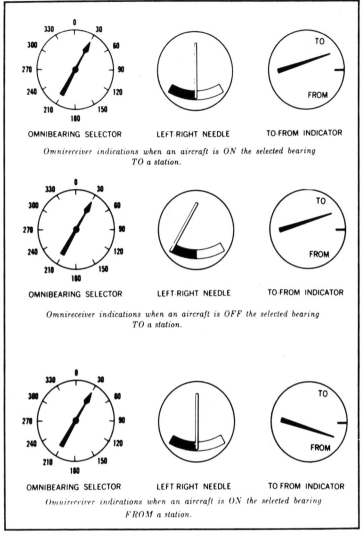

OMNIBEARING SELECTOR   LEFT-RIGHT NEEDLE   TO-FROM INDICATOR

Omnireceiver indications when an aircraft is ON the selected bearing TO a station.

OMNIBEARING SELECTOR   LEFT-RIGHT NEEDLE   TO-FROM INDICATOR

Omnireceiver indications when an aircraft is OFF the selected bearing TO a station.

OMNIBEARING SELECTOR   LEFT-RIGHT NEEDLE   TO-FROM INDICATOR

Omnireceiver indications when an aircraft is ON the selected bearing FROM a station.

Fig. 11-3. Omnireceiver indications.

VOR stations, assigned three-letter code identifications, are now being uprated to include voice identifications also (i.e.: Dallas VORTAC, -.. .- .-.., Dallas VORTAC, -.. .- .-.., etc.). VOR frequencies also are used by FSS personnel for weather broadcasts and other communications.

To operate your VOR receiver, tune in the VOR station as with other radio receivers and positively identify it, either by voice or

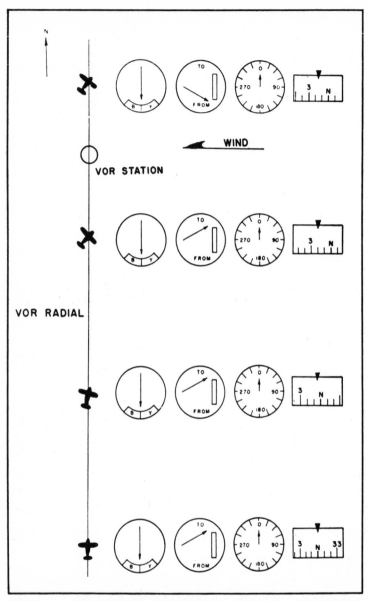

Fig. 11-4. How the aircraft compensates for a cross wind when flying the VOR. This drawing shows that when the pilot keeps the deviation indicator needle centered, he automatically compensates for a cross wind. Change of the TO-FROM indicator needle shows that this automatic wind-drift correction continues without regard to the location of the VOR station. Instruments shown are (left to right); deviation indicator, TO-FROM indicator, bearing selector, and compass. (courtesy FAA)

code, then manually rotate the omnibearing selector (OBS) until the LEFT-RIGHT needle centers. If the TO-FROM indicator reads FROM, rotate the course selector 180 degrees and get a TO reading when the LEFT-RIGHT needle centered (Figs. 11-1 through 11-5).

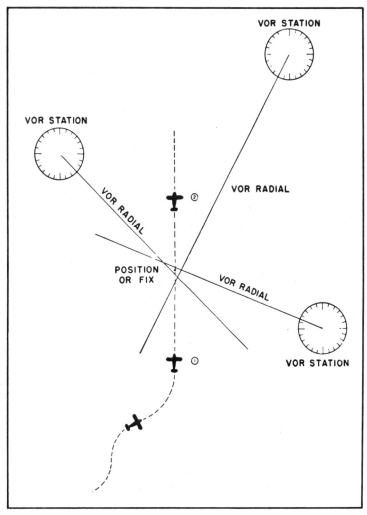

Fig. 11-5. Locating position when lost. When a pilot is in doubt about his position, VOR provides an easy and fast method of orientation. At point (1), the pilot takes up and holds a definite compass heading, tunes in and identifies three VOR stations, and notes location of each on chart. Between points (1) and (2), the pilot has obtained bearings of radials from three stations previously identified, notes time and on the chart plots lines or radials from stations. The center of the triangle formed by the radials is pilot's position at time of taking radials.

All you have to do then is turn to the heading that will hold the course shown to fly right over the station. Once past it, the TO-FROM indicator automatically flips to FROM, and if you wish to continue on the same track, simply keep the LEFT-RIGHT needle centered and away you go.

Another nice feature of VOR radio navigation is that you don't have to be flying toward or away from a VOR station (as you did with the old low frequency Adcock range beams) to remain oriented. By tuning in two or more VORS in the vicinity of your flight you can obtain the bearings of their radials; where they intersect is your position, or *fix*.

Radio Cross-Country

Let's take a cross-country trip, using radio navigation, and see how simple it really is.

You're sitting in the coffee shop at Wiley Post Airport, Oklahoma City, with your family looking over your Oklahoma City and Dallas Sectional charts and your Low Altitude En Route Radio Facility Chart. You've completed writing up your UFR Flight Plan, and on the backside your Pilot's Preflight Checklist.

You leave the waitress a tip, get a smile, and go to the telephone to file your Flight Plan (you've already checked the weather and figured ground speed). A Flight Specialist at the local FSS takes down the information from your Flight Plan and enters it into the teletype network to alert Flight Service Stations ahead that you're coming and when to expect you (Fig. 11-6).

DEPARTMENT OF TRANSPORTATION—FEDERAL AVIATION ADMINISTRATION
FLIGHT PLAN
Form Approved
OMB No. 04-R0072

1. TYPE	2. AIRCRAFT IDENTIFICATION	3. AIRCRAFT TYPE/ SPECIAL EQUIPMENT	4. TRUE AIRSPEED	5. DEPARTURE POINT	6. DEPARTURE TIME		7. CRUISING ALTITUDE
VFR					PROPOSED (Z)	ACTUAL (Z)	
IFR							
DVFR			KTS				

8. ROUTE OF FLIGHT

9. DESTINATION (Name of airport and city)	10. EST. TIME ENROUTE		11. REMARKS
	HOURS	MINUTES	

12. FUEL ON BOARD		13. ALTERNATE AIRPORT (S)	14. PILOT'S NAME, ADDRESS & TELEPHONE NUMBER & AIRCRAFT HOME BASE	15. NUMBER ABOARD
HOURS	MINUTES			

16. COLOR OF AIRCRAFT

CLOSE VFR FLIGHT PLAN WITH__________FSS ON ARRIVAL

Fig. 11-6. FAA flight plan form.

You thank the man and go to your airplane, parked at the Transient Area, and run your preflight inspection. Then you climb inside, put your Thermos bottle of coffee where you can reach it, turn on the master switch, tune your radio to 121.7 MHz (Ground Control) and speak conversationally into the mike:

"Wiley Post Ground Control, this is Cherokee Two Niner Three Four Kilo, transient area, ready to taxi, VFR Flight Plan to Dallas Love Field. Over."

Ground Control replies: "Cherokee Two Niner Three Four Kilo, Wiley Post Ground Control, runway One Seven. Wind one six zero at one zero. Altimeter two niner niner seven. Time one three zero zero zulu. Taxi north on ramp. Hold short runway One Seven."

You acknowledge: "Cherokee Three Four Kilo. Roger."

You release your brakes and taxi as instructed to the warmup area near the end of the runway, complete your preflight cockpit check, and switch both transmitter and receiver to the Tower frequency (120.65). You call:

"Wiley Post Tower, Cherokee Two Niner Three Four Kilo ready for takeoff."

The controller replies: "Cherokee Three Four Kilo, taxi into position and hold."

There is another craft just taking off, and the controller wants to space you safely behind him. Taking a good look to see that nobody is about to land (it's your neck!), you roll onto the runway, turn into the wind and wait.

Tower: "Cherokee Three Four Kilo, cleared for immediate takeoff."

You reach down and click the mike switch a couple of times to acknowledge, as there is no reason to sit there saying goodbyes; then you go. You head for the open sky to depart the airport traffic area, which extends to a radius of five miles around the field, then switch your radio to the Oklahoma City VOR frequency (115 MHz). You listen for the coded identification, and it comes: "---, -.-, -.-."

Now switch your VHF transmitter to the Oklahoma City FSS and report your actual time off to initiate your flight plan:

"Oklahoma City Radio, this is Cherokee Two Niner Three Four Kilo off at zero four, VFR Flight Plan Oklahoma City to Dallas, over."

The FSS man answers: "Cherokee Three Four Kilo, Roger, out."

Now you want to fly to the Oklahoma City VOR facility to start navigating, so you tune your course selector until the LEFT-

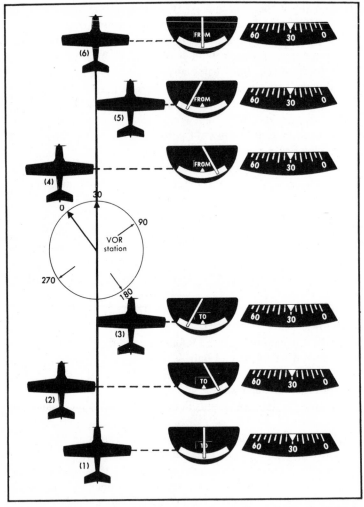

Fig. 11-7. VOR receiver indications for various positions of aircraft relative to desired course and VOR station. (courtesy FAA)

RIGHT needle centers (Fig. 11-7). Your Omni-Bearing Selector (OBS) reads 218 degrees, the TO-FROM indicator TO, so you turn so that magnetic heading and hold it until the TO-FROM reads FROM, the indication you've passed over the VOR station.

Your outbound radial along Victor Airway 163, which you've chosen to fly, is 145 degrees, so turn the OBS to 145 and commence a left turn until your deviation indicator tells you you're on course. You settle down, pour a cup of coffee, and marvel how easy it all is, flying high above the old Chisholm Trail with nary a rustler in sight.

If your ship is equipped with DME or TACAN, you don't even have to tune in on another VOR to get a position check; your airborne unit, called an interrogator, simply asks Oklahoma City VOR how far out you are on which radial. You flip the selector to Channel 97, and somewhere inside a bunch of dials begin to rotate, seeking a reply. Then, as if by magic, a red flag drops, dials spin, and there's the answer—040 nautical miles, like reading the odometer in your car! Another dial reads 324°, which is the reciprocal of your outbound bearing, 144°.

Halfway to the next VOR station at Ardmore, a small voice from the back seat says: "Daddy, how much further is it?"

You reply, of course, "Shuddup and lemme fly this thing!" Then you retune the VOR receiver to 116.7 MHz, the Ardmore frequency. To stay on Victor 163 Airway you want to fly into Ardmore on their 333° outbound radial, so you turn the TO-FROM indicator to TO and rotate the OBS to 153, the reciprocal of 333. If it happens you're to the right of that radial, the deviation needle will be over to the left. You simply turn *toward* the needle, maybe 30 degrees to the left of 153, or 123 degrees, until the needle centers. Then zap on in.

VOR stations are spaced roughly from 50 to 100 miles apart and provide a reception distance up to approximately 180 miles, depending on how high you're flying, because the signals are transmitted by line-of-sight.

You're flying high enough to read Ardmore loud and clear, which is nice because it's almost 1345 Zulu, time for a SIGMET (Significant Meteorological Information) broadcast covering a 400-mile radius.

The Ardmore FSS man begins.

"This is Ardmore Area Radio, time one-three four-five. Aviation weather, Ardmore, ceiling, five thousand broken, visibility four, light drizzle, fog. Temperature five-eight, dewpoint five-seven. Wind one seven zero degrees, six knots. Altimeter two niner niner two. This is Ardmore Area Radio."

From the back seat: "Daddy, I gotta . . ."

"Shuddup!" you growl. "I gotta figure!"

That character down at Ardmore has about spoiled your whole day—the barometer is falling, along with ceiling and visibility, and that temperature-dewpoint spread of only one degree means trouble. You flex your fingers; it won't be long before you'll be switching to the gauges, flying IFR.

The TO-FROM indicator flicks as you pass over Ardmore,

down under those building clouds. You turn your transmitter to 122.1 and call up the local FSS, leaving the receiver on 116.7, the Ardmore VOR frequency, which you need for navigation.

"Ardmore Radio, this is Cherokee Two Niner Three Four Kilo, over."

The reply comes via the VOR channel: "Cherokee Two Niner Three Four Kilo, this is Ardmore Radio. Go Ahead."

"Cherokee Three Four Kilo over Ardmore VOR at three five hundred, VFR Flight Plan Oklahoma City to Dallas, over."

"Cherokee Three Four Kilo, over Ardmore Radio Three five hundred, altimeter two niner niner two; Ardmore out."

Down below, the FSS man acknowledges your call, while upstairs you are listening to some PIREPS coming in over the Ardmore VOR—*Pilot Reports* from other airmen who recently passed down the airway ahead of you—mild turbulence and near-zero visibility ahead!

In your mind's eye you see the overall picture, as gigantic weather systems come together to clash in the heavenly battlefield, forming a front. This region, where hot, moist tropical air from the Gulf of Mexico onrushes against cold air from the northwest, has become known as *Tornado Alley*. It is here, along the Texas Panhandle, that ugly, black, snakelike whirlpools, highly charged with electricity, sometimes lash down to wreck havoc on the ground. They strike terror in the hearts of airmen who venture too close. They are not to be fooled with!

Once more you call up Ardmore Radio, requesting a change of flight plan from VFR to IFR, and the latest dope on the "big picture" of the weather system. You are relieved to learn that a squall line is not developing, but the weather definitely is dropping below VFR minimums. You are given the frequency to use to call the Fort Worth Air Route Traffic Control Center (ARTCC) direct.

Fort Worth is beyond your radio range, but you call them anyway to advise them you are changing to IFR. They hear you via the Remote Center Air Ground (RCAG) Communications Site nearby on a mountain top, a station that is tied into Fort Worth by land line (Fig. 11-8).

Down at the Fort Worth Center, your very good friend, the Air Route Surveillance Radar (ARSR) operator, already has his eye on you. You are a blip on his scope, and it comes to him via a Radar Microwave Link from a radar station at Oklahoma City, which has a 200-mile range.

The ARSR radar is a remarkable thing of more than four million

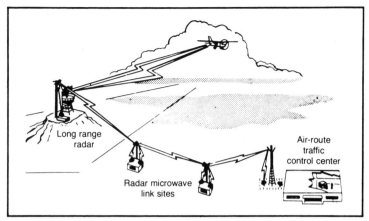

Fig. 11-8. Radar microwave link sites are located about 25 miles apart in the radar information link between the air route surveillance radar and the center.

watts of power that you can see your ship even inside rain clouds with a feature called *circular polarization*. Besides, it only responds to moving targets, so you won't be mistaken for a big hailstone.

What the radarman in Forth Worth sees is an enhanced blip with a comet-like tail shown on a *Radar Bright Display*, like watching a TV program in a lighted room.

Comfortable that big brother is watching you, you go on your way, letting neither rain nor sleet nor hail stop you.

"How much *longer*, daddy?" the voice from the back seat implores.

"Give him a cookie," you snarl. You are in trouble. Your omni receiver is acting up, and you're not too sure where you are. Your throat is dry, your hands clammy. But don't give up—call for a *DF steer!*

Fortunately for you, your Flight Information Manual shows, there's a Very High Frequency Direction Finder (VHF/DF) facility right at Love Field, where you're going. You call them up on their assigned frequency:

"Hello, ah, Love Field. This is, uh, I mean this is Cherokee Two Niner Three Four Kilo. Request homing, over."

You hide your tension, watching the rain splatter blowing off your windscreen, putting your whole faith in an unknown hero somewhere down there in the storm. He answers you:

"Cherokee Two Niner Three Four Kilo, this is Love Field. Transmit for homing, over."

Your heart beats faster now: "Love Field, this is Cherokee Three Four Kilo! AGH-H-H-H-H-H-H-H!" You do as the manual says and transmit a steady voice tone for ten seconds, until your face purples. You hope they got you, because you're out of breath! Then:

"Cherokee Three Four Kilo, this is Love Field. Course with zero wind one eight zero. Over."

You sigh with relief: "Love Field, Three Four Kilo, I read you, course one Eight zero, and thanks! Out."

Well, that confirms what you had thought, though you weren't sure—your DF course to Love is 180°; since you were on a course

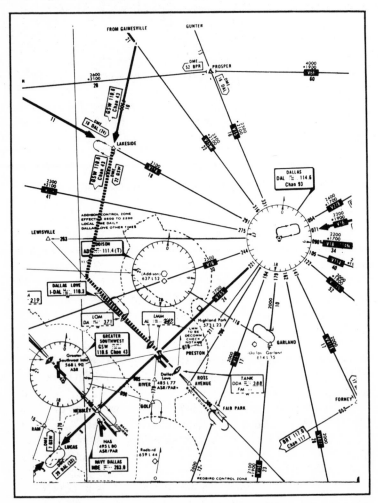

Fig. 11-9. Dallas terminal area ATC control (courtesy FAA)

of 150° to Dallas VOR, northeast of Love Field, the airport is due south of you (Figs. 11-9, 11-10).

You reach in the glove compartment and get out the cookie bag and pass one to the back seat. "We'll be home before you can eat that!" you grin, settling down to fly the gauges on home to the barn. The ARTCC guy at Fort Worth comes on the horn and tells you that you're being handed off to Radar Approach Control (RAPCON) at Love Field and would you please switch frequencies? The big eye watching you now is the Airport Surveillance Radar (ASR), with a 60-mile range; it will vector you down to where you can intercept the Instrument Landing System (ILS) to make your final approach. You're still in the soup, and you are thankful that the ASR has 400,000 watts of power keeping track of you, all snug in your cocoon of airspace—safer than on the freeway.

Your first contact with the ILS is the Localizer, a transmitter that sends out signals defining an extension of the runway centerline. Your receiver converts these signals into a needle reading much like your VOR indicator—in fact, many aircraft use the same instrument for both VOR and ILS (Figs. 11-11, 11-12).

The transmitting antenna you are tuned in on is at the far end of the active runway, and the man in the control tower can talk to you on the localizer frequency (in the range 108.10-111.95 MHz).

Once you settle down on the approach centerline, your next step is to watch for the Glide Slope Indicator needle to tell you when to start down to the runway. The Glide Angle, approximately a three-degree slope, extends out for about 25 miles from the transmitter, which is located near the approach end of the runway and about 600 feet to one side of the centerline.

A purple light flashes on the instrument panel—you've reached the outer marker, a 75-megacycle beacon between four and seven miles from the runway end. You check your altimeter; you're 1,200 feet above terrain . . . everything is green. You reset power to begin your letdown, about 475 feet per minute, a shade under 500 . . . speed 90 knots.

You peer ahead, straining, shifting your eyes from the ILS needles to the outside, back to the artificial horizon to keep your attitude right, back to the ILS needles. All is black out there, black as the inside of a catastrophe. A drizzle still splatters against the windscreen.

"Daddy! Can I have another cookie?"

Your knuckles whiten. You will attend to domestic problems later; now you're under real pressure. If you don't break through the

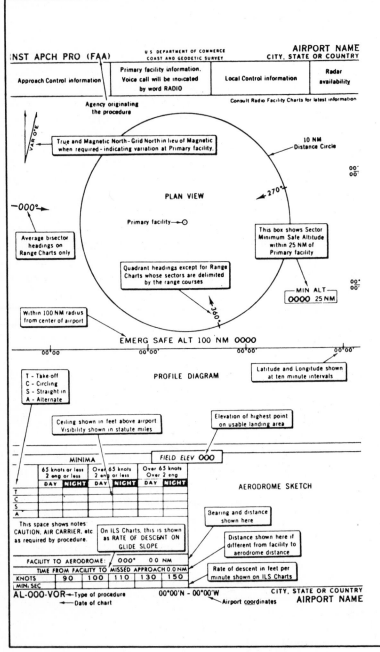

Fig. 11-10. Airport instrument approach procedure chart. (courtesy FAA)

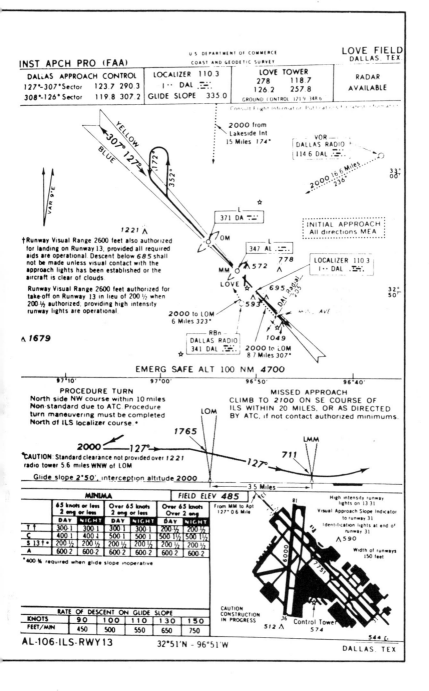

DALLAS APPROACH CONTROL	LOCALIZER 110.3	LOVE TOWER	RADAR
127°-307°Sector 123.7 290.3	I·· DAL ·⊡·	278 118.7	AVAILABLE
308°-126° Sector 119.8 307.2	GLIDE SLOPE 335.0	126.2 257.8	
		GROUND CONTROL 121.9 348.6	

Consult Flight Information Publications for latest information

YELLOW
307° 127°
BLUE
172°
352°

VAR 9°E

2000 from
Lakeside Int
15 Miles 174°

VOR
DALLAS RADIO
114.6 DAL ·⊡·

2000 16.6 Miles
236°

33°
00°

L
371 DA ·⊡·   ☆

INITIAL APPROACH
All directions MEA

1221 ∧

†Runway Visual Range 2600 feet also authorized
for landing on Runway 13, provided all required
aids are operational. Descent below 685 shall
not be made unless visual contact with the
approach lights has been established or the
aircraft is clear of clouds.

Runway Visual Range 2600 feet authorized for
take-off on Runway 13 in lieu of 200 ½ when
200 ½ authorized, providing high intensity
runway lights are operational.

7 OM

MM
LOVE

L
347 AL ·⊡·

∧ 572

LOCALIZER 110.3
I·· DAL ·⊡·

778
∧

695
DAL Radial 203°

32°
50°

∧ 1679

2000 to LOM
6 Miles 323°

593

RBn
DALLAS RADIO
341 DAL ·⊡·   ☆

MIL AVE

1049

2000 to LOM
8.7 Miles 307°

EMERG SAFE ALT 100 NM 4700

97°10'   97°00'   96°50'   96°40'

PROCEDURE TURN
North side NW course within 10 miles
Non-standard due to ATC Procedure
turn maneuvering must be completed
North of ILS localizer course. •

MISSED APPROACH
CLIMB TO 2100 ON SE COURSE OF
ILS WITHIN 20 MILES, OR AS DIRECTED
BY ATC, if not contact authorized minimums.

LOM

2000 —127°→

1765

127°→

LMM

711

•CAUTION: Standard clearance not provided over 1221
radio tower 5.6 miles WNW of LOM

Glide slope 2°50', interception altitude 2000

3.5 Miles

MINIMA						FIELD ELEV 485		From MM to Apt	High intensity runway

	65 knots or less 2 eng or less		Over 65 knots 2 eng or less		Over 65 knots Over 2 eng	
	DAY	NIGHT	DAY	NIGHT	DAY	NIGHT
T †	300-1	300-1	300-1	300-1	200-½	200-½
C	400-1	400-1	500-1	500-1	500-1½	500-1½
S 13 †•	200-½	200-½	200-½	200-½	200-½	200-½
A	600-2	600-2	600-2	600-2	600-2	600-2

•400-¾ required when glide slope inoperative

From MM to Apt
127° 0.6 Mile

High intensity runway
lights on 13-31
Visual Approach Slope Indicator
to runway 31
Identification lights at end of
runway 31
∧ 590
Width of runways
150 feet

CAUTION
CONSTRUCTION
IN PROGRESS

Control Tower
512 ∧   574

	RATE OF DESCENT ON GLIDE SLOPE				
KNOTS	90	100	110	130	150
FEET/MIN	450	500	550	650	750

544 ⌐

AL-106-ILS-RWY13   32°51'N - 96°51'W   DALLAS. TEX

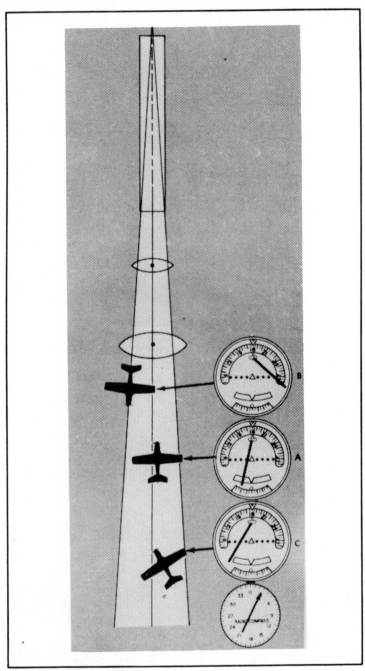

Fig. 11-11. Localizer receiver indications. (courtesy FAA)

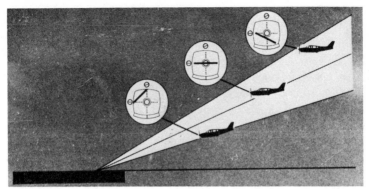

Fig. 11-12. What your glideslope indicator tells you—above, on, or below glide-slope. (courtesy FAA)

overcast soon, it means a go-around and more sweat. You pick up the mike and call Approach Control:

"Cherkokee Two Niner Three Four Kilo, request Precision Approach control monitor. Over."

The man replies: "Cherkokee Three Four Kilo, affirmative."

No need for long conversation here—everybody's busy getting you down through the soup, and with Precision Approach radar

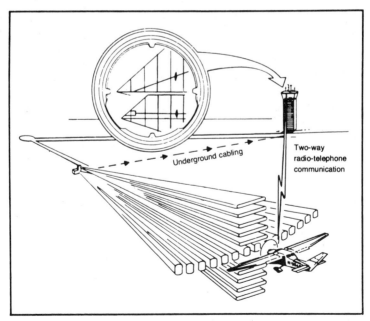

Fig. 11-13. Precision approach radar with only communication equipment. (courtesy FAA)

Fig. 11-14. Glideslope indicator—needles centered.

(PAR) you know the radar man can confirm your altitude, in case your altimeter is off (Figs. 11-13, 11-14).

No ground yet . . . but your ILS needles have you lined up on course, dead center, dead on the glide slope. An amber light on your panel glows—the inner marker. You are 3,500 feet from the runway, your altimeter says you're at 200 feet, and the PAR man says: "Everything looks good!"

Suddenly, out of the goop ahead, a running ball of light grabs your attention—the flashing strobe lights of the Approach Lighting System, pointing the way to the runway! You're through the overcast! You're flying VFR once more, crossing the 1,000-foot wing bar, and the threshold green lights spell out *welcome*!

Farther down the runway you see other lights, red and white and green—the VASI lights, Visual Approach Slope Indicators (Fig. 11-15). Too low, they look red; too high, white; dead on, green!

The *screech* of tires on concrete brings you back home again, and even if the airport is socked in with fog, there's a man in the tower watching you on an Airport Surface Detection Equipment (ASDE) radar to tell you where to taxi.

All this is the result of a miracle of electronics, and you are so happy that you go to the telephone booth (after closing your flight plan) to call home. You fumble and fumble. No dime.

"What in hell are we coming to?" you grumble.

A small voice behind you makes a last, plaintive sob: "Daddy,

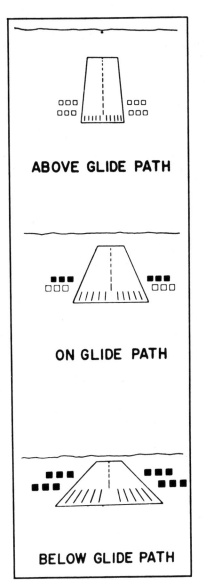

ABOVE GLIDE PATH

ON GLIDE PATH

BELOW GLIDE PATH

Fig. 11-15. How pilot sees VASI (Visual *A*pproach *S*lope *I*ndicator) lights. (courtesy FAA)

can I have just one more cookie?"

Your tensions ease, you give the kid a big kiss and start home, glad to be alive.

ADF

Flying in Canada, Mexico, and sections of the western United

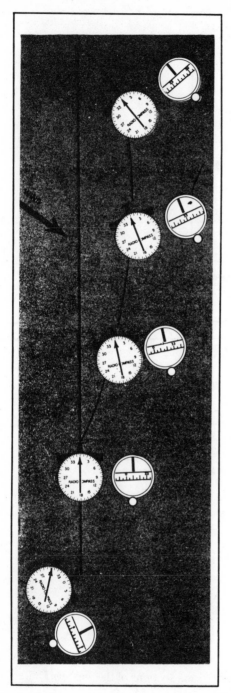

Fig. 11-16. ADF homing in crosswind. (courtesy FAA)

States takes you beyond the range of VOR stations. Thus there's a good reason to have on board a nice Automatic Direction Finder (ADF) set, which provides you with five benefits:

- ☐ It serves as a backup navigation system in the event of VHF equipment failure.
- ☐ It can be used to monitor your position en route and provide data for plotting fixes.
- ☐ It makes a fine navigation system in areas and at low altitudes where VOR "line-of-sight" signals are unreliable.
- ☐ Sitting on the ground, where you can't tune in a VOR station, the ADF can be used to monitor weather reports (and football games).
- ☐ It can also supply auxiliary and standby navigation information when flying instrument approaches.

When used in navigation, the ADF needle points TO the transmitting station regardless of your heading or position. The relative bearing thus indicated is the angular relationship between aircraft heading and the station measured clockwise from the nose of the aircraft.

To home on a radio station (even a commercial broadcast variety), turn left or right to a heading that will zero the ADF needle (Fig. 11-16).

As you close in on the station you'll be making more frequent heading corrections to keep the needle on zero, particularly in a strong crosswind. Passage over the station is indicated by a 180° reversal of the ADF needle to the tail position.

However, an ADF approach executed without other airborne navigational equipment takes some doing, both in basic flying and use of nav instruments, so make sure you're familiar with courses, altitudes, and procedure details before attempting one.

Chapter 12
Radar Flying the Easy Way

When can you legally fly IFR in controlled airspace without filing an IFR flight plan?

What is a pop-up?

Must a VFR-rated pilot follow assigned radar vectors that will take him into IFR conditions?

When should an arriving VFR pilot contact Approach Control for Stage II or Stage III Radar Advisory and Sequencing Service?

Where do you look up the frequency to contact ATC for Stage II or Stage III service?

These are typical questions asked today by competent, rated airmen flying transponder-equipped personal aircraft, but frequently hesitant about using radar service simply because they're unfamiliar with it except in a general way. As a result, too many pilots sit on their hands waiting for the weather to break instead of doing the simple and obvious thing—calling for radar service.

For example I was recently flying southbound from Fresno, California, on a VFR flight plan, following Victor 23 Airway after a weather briefing that showed the Los Angeles Basin clear and four miles in late afternoon. My destination was an uncontrolled field in San Fernando Valley, Whiteman Airport, and from past experience I knew the weather could drop below minimums in a matter of minutes.

At Bakersfield I rechecked the weather by calling Flight Watch on 122.0 and found conditions still VFR at nearby Van Nuys and

Burbank, the closest reporting points, so I proceeded. Then, some 25 miles out from the Van Nuys VOR, I tuned to that field's ATIS (Automatic Terminal Information Service) and was told the visibility had dropped to three miles with an 1100-foot ceiling. I realized the usual evening stratus deck had moved in, and there I was in growing darkness, VFR on top (Fig. 12-1).

At this point I called Burbank Approach Control on 120.4 and requested Stage II radar service, which provides safe sequencing between other traffic. I found that frequency in Part 3 of *AIM*, gave my aircraft identification, position (TWINE Intersection) and altitude (5000 feet), then asked for radar vectors to Whiteman. I also advised I had ATIS from Van Nuys and had instrument capability.

With this input, the controller assigned a beacon number to squawk on my transponder, gave me an inbound heading to steer, and requested that I remain VFR until further notice. He was able to punch the data into the Burbank station's computer, which did two things. It displayed the data on my blip on the ARTS-III automated radarscope, and passed it along to Los Angeles Center's computer at Palmdale via an electronic interface.

This process in effect "bored a hole in the sky" for my safe progress, ensuring proper sequencing with all other identified VFR and IFR traffic in the radar environment. I then received clearance to descend through the cloud deck to VFR conditions below, and when I advised that I had Whiteman in sight, radar service was terminated. I had, in fact, been cleared for a modified IFR approach while flying under a VFR flight plan.

Fig. 12-1. Evening stratus clouds cover Los Angeles Basin.

A bonus of the system is that there was no need to load up the Center's computer with extraneous flight-plan data when all I needed was local radar service. ATC could care less about the color of my aircraft or where I lived, but if I had been low on fuel, they would like to know precisely *how* low. The whole system depends on information input. If you don't tell them your problems, they won't know what answers to give.

This is why it's a good idea to make your initial callup far enough out, when arriving in a terminal traffic area, for the controller to punch in your data. That way you're set up for an efficient, safe approach without having to be assigned to a holding pattern.

Stage II and Stage III Service

When the FAA first developed the "Stage" services, two basic radar features were identified as Stage I: traffic advisories and limited vectoring. Today these basic services are included in Stage II or III, and Stage I has been dropped. When you call up ATC to request radar service for either Stage II or III, after receiving ATIS information, tell them quickly: "Have numbers." And if you don't need radar service, say "negative Stage II or III," whichever applies.

When you make initial contact 25 miles out, Approach Control will assume you want radar service unless you tell them. Incidentally, Stage III radar service is provided entering airspace defined as a Terminal Radar Service Area (TRSA) but again is not mandatory. It's a service, not a requirement, but it's a good way to fly!

If Approach Control is not expecting your arrival, you're known as a pop-up. This tells the controller that you probably don't have IFR approach plates aboard, so instead of giving you, say a Radar-1 IFR approach, they'll simply give you headings (vectors) and descent instructions to follow, meanwhile watching your progress on the radarscope.

Transponders

Transponders have been around for quite a while, having been developed during World War II, and initially worked on a two-dimensional format. Today, when flying into specific Stage III TRSAs and TCAs (Terminal Control Areas), three-dimensional Mode C transponders are required, to give ATC automatic altitude information as well.

An example of how useful en route radar service can be on a normal VFR cross-country flight was the time I was eastbound across Ontario, Canada, headed for Niagara Falls Airport via Sault Saint Marie and the Lake Huron island chain. I was dogging it behind a slow-moving cold front whose towering cumulus formations reached for miles directly across my route.

All went well until I approached the westerly tip of Lake Ontario, where the cold front stalled, becoming a solid occlusion. There I was, VFR on top of rapidly deteriorating weather. It was time for a decision: Should I do a 180 and return to Wiarton, Ontario, my last fuel stop, and risk finding the weather below minimums there too, or file IFR and press on?

Though IFR-rated, I had only marginal IFR instruments aboard—no glide slope and only a single NAV/COM radio. But I had recently installed a Narco AT-50 transponder, and as yet had had no reason to use it. I called Buffalo Radar, told them my situation, and requested vectors to descend to the vicinity of Niagara Falls Airport.

"November Thirteen Delta Delta," radar replied, "squawk one-two-zero-zero."

I pressed the transponder's ident button, which amplified my blip on the Buffalo radarscope, and within moments I was under positive control through IFR weather, though still under a VFR flight plan. Descending below 2000 feet, under a balloon ceiling and with visibility barely three miles, my radar contact ended, and I simply followed my last assigned heading until the airport was in sight.

Incidentally, the term, "squawk" comes from the World War II era, when military transponders were called "parrots" in secret wartime jargon, and they replied to coded electronic messages just as the raspy-voiced green birds do.

Stage II radar can be mighty helpful to a VFR pilot with no IFR capability who finds himself flying in haze, barely legal and maybe lost, a confused transient trying to find an airport far from home base (Fig. 12-2). He simply contacts the proper radar facility, tells them his plight, and gratefully accept vectors to follow, with the Big Eye watching him.

There's a problem, though. The Big Eye may not be able to see clouds so well, and your assigned vector may take you right into the top of a towering cumulus dead ahead. So do not follow directions—instead, advise the controller of the problem and ask for vectors around the weather, not through it.

Some pilots are reluctant to contact ATC for Stage II or Stage III service due to their unfamiliarity with terms used. The control-

Fig. 12-2. A VFR pilot can easily become lost in Los Angeles smog.

ler may be a rapid-fire talker, so just pick up the mike and request: "Say again, *slowly*, please?" The heart of the system is close communication between pilot and controller, so don't be afraid to ask.

If you're equipped with a Mode C transponder, however, the thing responds automatically, displaying position, altitude, ground speed, etc., on your radar blip in symbols and alphanumerics. Still more sophisticated radar equipment lies ahead under the National Airspace System Plan—a new secondary radar system known as Mode S.

Mode S, like the Air Traffic Control Radar Beacon System (ATCRBS) it will replace, obtains information on aircraft by querying transponders in these aircraft, reading out the coded replies, and presenting these data on radar displays. Mode S can also interrogate aircraft individually to circumvent current interference problems. It also will provide automatic air-ground data link communications, providing direct pilot access to automated data bases, using a cockpit keyboard. This system is expected to be in use by 1990 for all aircraft flying above 12,500 MSL.

Flight Watch

We touched only briefly on another vital radar service for cross-country pilots—Enroute Flight Advisory Service (EFAS), known to airmen as simply *Flight Watch*. Once airborne, after a preflight weather briefing, the cross-country pilot is concerned about the possibility of deteriorating weather ahead. He simply tunes his radio to 122.0 uses the name of the controlling Flight Service Station (or simply say "Flight Watch" if controlling FSS is unknown), giving identification, position, and request for desired information.

Today some 44 EFAS stations have been implemented across the United States, each of which is in touch with the National Weather Service's network of 56 weather radars. Selected FSS also provide transcribed weather broadcasts, including the highly important Pireps (Pilot Reports) that provide up-to-date local weather data.

EFAS is not to be used for filing flight plans or to request the type of weather reports you'd normally get during your preflight briefing, but the EFAS specialist will provide you with any data you really need, such as altimeter settings at your destination airport, which can change abruptly.

I frequently use the Los Angeles Flight Watch flying into Los Angeles Basin after a campout on the Mojave Desert, where contact with a FSS or NWS station is not available. Leaving the Death Valley area, I call 'em up on 122.0, 80 miles out from Daggett; based on the Los Angeles weather report, I'll know whether I'll be making an ILS or VFR approach.

At the same time I pass along a PIREP on the weather picture I've encountered—turbulence, cloud cover, visibility, and so forth. I report my location and aircraft type to help them evaluate my turbulence report—in my Cessna, what I consider moderate turbulence might seem severe to a Cub driver. Cloud ceilings are reported in MSL rather than AGL values.

There are other ways to evaluate en route weather—namely, by using an airborne weather radar system. Today they are priced under $10,000, and are becoming more widely used in both business and private aircraft (Fig. 12-3).

Weather radar sets normally reflect areas of moderate to heavy precipitation, though radar does not detect turbulence directly. Experience has shown that frequency and severity of turbulence generally increases with the radar reflectivity closely associated with areas of highest liquid water content in a storm. *AIM* warns that "No flight path through an area of strong or very strong radar echos separated by 20-30 miles or less may be considered free of severe turbulence."

Airborne radar units feature color-coding, with red areas indicating regions of heaviest moisture concentration. Their antennae can be tilted in flight to monitor towering nimbocumulus formations, and normally are adjusted so that ground clutter returns barely disappear from the screen.

As effective as these radar services may be today, the trend of high-tech development in avionics spells the doom of some and opens the sky for intriguing new gadgets to make flying easier and safer. By the year 2000 the current en route primary radar service will be phased out, though terminal primary radar with a weather channel will be continued. A new weather radar will offer improved weather services for the ATC, and the Mode S transponder already mentioned will form the heart of a new full collision-avoidance method called TCAS (Traffic Alert and Collision-Avoidance System).

RNAV

Back in 1973, the FAA proposed a sweeping new program calling for a total overhaul of the airspace structure above 18,000 feet, and a redesign of high and medium density terminal areas for the exclusive use of a thing called RNAV—Area Navigation (Fig. 12-4). The changes would be as sweeping as those of the early 1950s, when the old network of low-frequency, four-course radio range transmitters (Adcock ranges) were replaced by the VOR system.

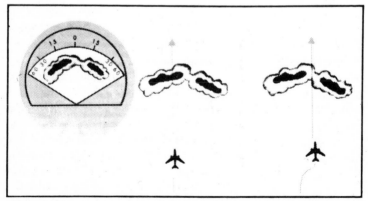

Fig. 12-3. Airborne weather radar spots storm cells. (courtesy FAA)

Fig. 12-4. RNAV (area navigation) sets set up "phantom" waypoints.

RNAV is based on use of small airborne computers that permit pilots to fly any desired course by extrapolating signals from ground navigation aids. This frees them from the necessity of flying directly between ground stations. The following year the FAA proposed upgrading RNAV to include three-dimensional capability to give pilots desired ascent and descent angles to get in and out of airport areas. This is similar to the ILS approach system but with much greater range.

The RNAV program ran into a snag out west, when the FAA's Western Region set out to establish three low-altitude RNAV routes between Los Angeles and San Francisco. The problem was they couldn't get continuous coverage along the route due to the presence of high mountains, so all they could do was simply parallel existing airways.

However, at high altitudes, and flying overseas, RNAV would permit flying a great circle route, providing you stayed above 18,000 feet—ATC wasn't set up to handle such a system. So today, except for a few high-altitude Alaskan routes, a United States RNAV system does not exist, though provisions are retained for designation of RNAV routes in the future, should they be required. In absence of established RNAV routes, random RNAV routes may be filed, though ATC monitoring will be a requirement.

LORAN-C

Typical of the RNAV systems are VOR/DME RNAV, INS, OMEGA, and LORAN-C. I remember flying a DC-3 ferry flight from Miami to Manila using an early LORAN-A unit as big as a kitchen cabinet, though today they're relatively small.

LORAN stands for *LO*ng *RA*ange *N*avigation, and is an electronic system using shore-based radio transmitters and airborne (also shipboard) receivers. The system was developed during World War II as LORAN-A by the Radiation Laboratory of the Massachusetts Institute of Technology, and subsequently operated by the U.S. Coast Guard to fulfill wartime needs. By 1971 a total of 83 stations were in service.

Meanwhile, in the late 1950s and early 1960s, the Department of Defense instituted a program designed to develop a new generation of radio navigation aids, and the result was LORAN-C. Today there are 13 LORAN-C chains operating in the United States and overseas.

Each LORAN-C chain consists of three to five land-based transmitting stations separated by several hundred miles. One station is designated the master station (M), the others secondary stations—Whiskey (W), X-ray (X), Yankee (Y) and Zulu (Z). Signals transmitted from the secondaries are synchronized with the master signal, at precise time intervals. The on-board LORAN-C receiver measures the slight difference in time that it takes for these pulsed signals to reach the aircraft, measured in microseconds (millionths of a second). The time difference is plotted on a LORAN-C chart as a line of position. By plotting such lines with others taken from other secondary (slave) stations, a position fix is quickly obtained.

Unlike VHR RNAV systems, LORAN-C permits the pilot to establish waypoints at locations not feasible using VHF. Tuning is automatic, once you have properly tuned in the desired network or chain of stations.

NAVSTAR GPS

Far more sophisticated than LORAN-C is the latest navigation system to come along, NAVSTAR GPS. It's similar in function to LORAN-C, but instead of utilizing ground-based transmitters, it operates on space satellite linkups (Fig. 12-5).

NAVSTAR GPS is an acronym for *NAV*igation *S*atellite *T*iming *A*nd *R*anging *G*lobal *P*ositioning *S*ystem. Already six NAVSTAR satellites are in orbit, funded by the Department of Defense. In May 1983 a $1.2 billion USAF contract was awarded to

Rockwell International for an additional 28 NAVSTAR satellites to be launched from space shuttles. Of the 28 satellites, only 18 will be in orbit at one time, the others used for spares or replacements.

NAVSTAR GPS was designed to solve a problem of other long-range radio navigation systems—high-frequency radio waves

Fig. 12-5. Lightweight Collins NAVSTAR GPS unit weighs 15 pounds. (courtesy Rockwell International).

provide accurate navigation signals but are limited to line-of-sight coverage, while lower frequencies hugged the Earth or were reflected from the ionosphere.

GPS was conceived in the 1960s as the USAF's highly classified System 612B, and subsequently became a joint-services development. With signals from 24 satellites orbiting the Earth at an altitude of 10,898 nautical miles, GPS would provide 24-hour, all-weather, worldwide, three-dimensional (latitude-longitude-altitude) fixing with accuracies of about 60 feet.

A GPS position is similar to a LORAN-C fix from several widely-separated transmitters on Earth. Each satellite continuously broadcasts its position in space, along with its ident and other data.

The user's GPS receiver monitors these signals and after deciding which four satellites of those visible above the horizon provide the best fix geometry, measures the length of time it takes for the individual signals to arrive from three of them. Multiplying these by the speed of radio waves (the same as the speed of light) gives three ranges to work from. The three ranges are automatically used to calculate the receiver's position in space relative to the center of the Earth, to provide latitude, longitude, and height above sea level.

Such triangulation requires extreme precision in time measurement, in billionths of a second, and here's where the fourth satellite comes in. Each satellite carries a super-accurate atomic clock, and as your receiver is tracking three satellites for ranging, a quartz clock in the set synchronizes itself with the atomic clock in the fourth satellite.

A GPS set already is being used in an evaluation program by the USAF, utilizing a low-cost prototype "Z" set by Magnavox Corporation, costing under $30,000 and weighing 33 pounds, aboard a Sikorsky S-76 helicopter.

Chapter 13
Aerobatic Flying

In recent years, more and more general aviation and airline pilots alike are turning to aerobatic flying both for recreation and to sharpen their skills. Precision aerobatics are not to be confused with airshow stunt flying (Fig. 13-1), which may include such wild maneuvers as the *lomcevak*, which we'll tell you about later.

According to the FAA, aerobatic (or acrobatic) maneuvers are intentional maneuvers involving an abrupt change in an aircraft's attitude, an abnormal attitude, or abnormal acceleration, not necessary for normal flight.

The FAR law also says you're supposed to wear a parachute to bank more than 60 degrees or tilt your nose up or down more than 30 degrees relative to the horizon, except on check rides with a CFI or an airline transport pilot.

Definitions don't mean much, however, unless you know what they are based on. To the FAA, limitations are imposed to keep you from exceeding design load factors of your aircraft and bending something out of shape.

The mark of a good aerobatic pilot is not how much punishment he can give himself or his aircraft, but how smoothly and gently he can execute complex maneuvers, getting the most out of his ship. Art Scholl (Fig. 13-2), a member of the United States Aerobatic Team for two successive competitions behind the Iron Curtain, flies his Canadian-built Chipmunk and modified Super Chipmunk with such ease one could close his eyes riding with him and hardly know

Fig. 13-1. Red Devils aerobatic team provides thrills at summer airshows put on by Experimental Aircraft Association at Oshkosh, Wisconsin.

he was not flying straight and level. In a jocular mood, Art once insisted to an FAA inspector that flying along a runway inverted to pick up a handkerchief with his rudder (Fig. 13-3) was *not* an aerobatic maneuver. "I was flying straight and level," he said with a straight and level face. "And besides, the Chipmunk is approved for continuous inverted flight!

Why Aerobatics?

One authority on aerobatic flying is the great Duane Cole, a member of the Cole Brothers Air Shows and author of *Roll Around a*

Fig. 13-2. Art Scholl runs School of Aerobatics.

Point. This is an excellent manual covering ten hourly lessons and describing the peculiar script of the *sistema aerocriptografico Aresti*—a system of diagramming aerobatic maneuvers designed by Count Jose L. Aresti of Spain, and used in all national and international competitions.

Watching Duane Cole execute square outside loops, outside-inside Cuban eights, and freestyle routines in his clipped-wing Taylorcraft is an unforgettable experience; you know you are watching a master at work.

Duane offers this bit of good advice to the pilot who wants to take up aerobatics:

"An aerobatic pilot learns to fly with his head out of the cockpit. He must be cognizant of his position relative to the ground, the attitude of his airplane relative to the horizon, and above all, the position of any traffic in the area. The division of attention required to fly a sequence of maneuvers is invaluable in strengthening your perceptibility, and a result, enhance your chances of survival in this era of hurrying and scurrying whether you are flying, driving, or even walking."

In addition to these awarenesses of attitude and environment, the good aerobatic flyer develops a sense of feel, not unlike the sensory perceptivity of the old seat-of-the-pants pilot, that tells him

Fig. 13-3. Art Scholl flies Super Chipmunk inverted for pickup.

is about to exceed his critical angle of attack, whether flying rightside-up or upside-down.

Aerobatic Maneuvers

Two excellent training maneuvers, the chandelle and the lazy eight, although not categorically aerobatic are the most lovely and exhilerating of all basic training flying routines with which to develop subconscious feel, planning, orientation, coordination, and speed sense. You just can't do a lazy eight mechanically; the control pressures required for constantly changing coordination are never exactly the same.

No two pilots seem to agree on just how a chandelle should be flown, though by definition it is supposed to be an altitude-gaining maneuver with a complete reversal of direction. The preferred FAA method regards the chandelle as a 180-degree change of direction with maximum altitude gain, coordinated flight, and completion of the turn with wings level and airspeed at V_{mc} (minimum control speed).

Enter the chandelle, says the FAA, at normal cruise or maneuver speed (which is less) and not above the manufacturer's recommendation.

From straight and level flight, establish a 30-degree bank (if you dive for speed, return to level flight before banking), then smoothly apply up-elevator pressure to increase pitch attitude to the 90-degree turn point.

From there until you complete the 180-degree turn, gradually decrease the bank while maintaining constant pitch attitude, arriving at the 180-degree point with wings level, airspeed just above a stall. During the turn, power is added as soon as the airspeed starts to deteriorate.

Thus, in the first half of the chandelle you are holding constant bank and changing pitch; in the last half, constant pitch and changing bank.

The lazy eight is not related to other types of figure eight maneuvers, and is so named only because of the figure apparently drawn on the horizon line when viewed sideways.

In executing this beautiful maneuver, the pilot at no time flies straight and level; he is constantly rolling from one bank to another, snaking his way across the sky in a series of 180-degree climbing and diving turns.

Again, the FAA prefers that these be done with a shallow, 30-degree bank. This is actually harder to do than a lazy eight with 60-degree banks, requiring far more skill in coordination and speed control.

In this maneuver, the high point of the first climbing turn is reached 45 degrees from the point of entry, instead of 90 degrees as in the chandelle. At the 90-degree point, the bank is at its maximum, the airspeed at its minimum, and the pitch attitude passing through level flight attitude to continue in a diving turn with pitch decreasing, airspeed increasing, and bank gradually shallowing.

The lowest pitch attitude is reached 45 degrees past the halfway point, and from there the bank gradually shallows and airspeed increases as the nose crosses the horizon with wings level to commence another climbing turn in the opposite direction.

Most pilots do lazy eights with a constant power setting, although they may be executed with changing power or no power at all, so long as the loops of the eight are symmetrical. A common error is to use the propeller spinner for a reference point in crossing the horizon at the 90-degree point. Instead, select a point directly ahead of your eye level for a reference, or you'll gain altitude.

As an introduction to inverted flight, the simple loop can be performed so smoothly that you are held firmly in your seat all the way around with approximately one G (gravity force), although it requires a nice bit of elevator-throttle coordination and use of right rudder to counteract the *P-factor* (propeller torque) as power is added and the airspeed slows down toward the top of the maneuver. A well-executed loop appears from the side as a perfect circle, though small, underpowered trainers achieve a more egg-shaped pattern.

In aerobatic flight schools, you will progress from fundamental maneuvers such as those just described to more complex ones like the beautiful clover leaf and Cuban eight, high-speed stall maneuvers like the snap roll, Immelman turns, barrel rolls, and hammerhead stalls.

To Art Scholl, the lomcevak is the most fun of all, a wild, gyrating, tumbling thing you have to see to believe. Originated in Czechoslovakia in the early 1960s, the lomcevak gets its name from a Czech word roughly translated as: "I've had too much slivovitz brandy and all my gyros have tumbled!"

To enter the maneuver, the pilot rolls inverted from level flight and pushes forward on the stick into a vertical climb at full throttle. At the top of the climb he executes an outside snaproll with full forward elevator, full left rudder, and full right aileron. The climb stops abruptly, and suddenly the nose is swinging wildly around the horizon through 360 degrees, like a gyroscopic top gone wild. The nose suddenly drops as gyro forces, acting in a direction 90 degrees from point of application, change the axis of rotation from vertical to horizontal. With a gasping plunge the aircraft tumbles end over end, again changing the axis of rotation. If left alone, the pilot can get out through an inverted spin, although Scholl likes to split-S out in a swooping dive.*

Manufacturers of general aviation aircraft today recognize the growing interest in aerobatic flying and are adding specialized new aerobatic ships to their production lines of utility category aircraft.

Among the first was the Champion 7ECA Citabria (airbatic spelled *sdrawkcab!*), first flown in 1962 as a two-seater tandem monoplane (Fig. 13-4). A fully aerobatic model, the 7KCA, appeared later with clipped wings and 150-horsepower Lycoming 0-320 engine.

Beech Aircraft Corporation (which long has catered almost exclusively to the business flier with sleek, fast and comfortable Bonanzas, King Airs, Queen Airs, and Barons) first began building trainers with aerobatic capabilities in 1948 with their two-place tandem T-34 Mentor. In 1968 they added to their line the versatile

---

*Actually, this is a variation of the true lomcevak—Ed.

Fig. 13-4. Champion Citabria (airbatic spelled backward).

two-place Musketeer Sport III, powered with a 150-horsepower Lycoming.

On the used plane market, of course, is a wide range of World War II primary trainers, such as the Ryan PT-22, the Stearman PT-17, and the Fairchild PT-19, the Vultee BT-13 basic trainer, and the North American T-6 advanced trainer. Smaller craft like the all-metal Luscombe 8E and the Monocoupe 90A may also be flown aerobatically.

There are, in fact, a plane for nearly every purpose and pocketbook, nearly two thousand accredited flying schools, and an open sky beckoning to you to join the half million pilots now making up America's civilian air strength.

There are also hot air balloons, gyroplanes, helicopters, and gliders—and in more than one backyard, inventors are still trying to defy gravity with man-powered ornithopters!

If you've read this far, your interest shows that you are a good potential pilot, for enthusiasm is about all you really need! So come on up! The air's fine!

Index

Index